# Lecture Notes in Computer Science

Edited by G. Goos and J. Hartmanis

212

# Interval Mathematics 1985

Proceedings of the International Symposium
Freiburg i. Br., Federal Republic of Germany
September 23–26, 1985

Edited by K. Nickel



Springer-Verlag
Berlin Heidelberg New York Tokyo

**Editor**

Karl Nickel
Albert-Ludwigs-Universität, Institut für Angewandte Mathematik
Hermann-Herder-Straße 10, D-7800 Freiburg i. Br.



CR Subject Classifications (1985): G.1.0, G.1.1, G.1.2, G.1.3, G.1.5, G.1.6, G.1.7, G.1.9

ISBN 3-540-16437-5 Springer-Verlag Berlin Heidelberg New York Tokyo
ISBN 0-387-16437-5 Springer-Verlag New York Heidelberg Berlin Tokyo



Printing and binding: Beltz Offsetdruck, Hemsbach/Bergstr.
2145/3140-543210

# Preface

The present volume contains the proceedings of an International Symposium on Interval Mathematics held in Freiburg i.Br./Germany from September 23 to 26, 1985.

The last two such international symposia took place in Karlsruhe/Germany in 1975 and also in Freiburg i.Br./Germany in 1980. The attendees of the present symposium came from 12 countries, namely from: Austria, Brazil, Bulgaria, China, Czechoslovakia, Japan, Poland, Spain, United Kingdom, USA, Yugoslavia and from West Germany.

By several reasons not all the papers presented at the Symposium could be included in this volume. The papers of Rohn (Czechoslovakia) and of Xu (China) have been added, although both where unable to come to Freiburg, by last minute reasons.

The International Symposium 1985 was made possible by grants of:

IBM Deutschland,
Gödecke AG / Freiburg i.Br.,
Verband der Freunde der Universität Freiburg i.Br.,
Rektorat der Universität Freiburg i.Br.,
Gesellschaft für Angewandte Mathematik und Mechanik (GAMM).

The GAMM-Fachausschuß für Intervall-Mathematik served as organizing committee. The success of this conference was due to hard work of many members of the Institut für Angewandte Mathematik der Universität Freiburg, especially to Dr. Garloff, to the institute secretary, Mrs. I. Kasper and to the two symposium secretaries, Mrs. H. Sturm and Mrs. L. Wüst. The editor wishes to express his gratitude to all of these, to the speakers and to the staff of Springer-Verlag.

Karl Nickel

# Contents

# INTERPOLATION OF AN INTERVAL-VALUED FUNCTION FOR ARBITRARILY DISTRIBUTED NODES

H. Engels and D. an May
Technical University of Aachen

## 1. Introduction and survey

The following problem is repeatedly posed in practice:
Given a set of $N$ points $(x_i,y_i) \in \mathbb{R}^2$, and intervals $I_i = [u_i,o_i]$, $i=1(1)N$, such that for a function $f(x,y)$ to be interpolated the function values $f_i = f(x_i,y_i)$ satisfy $f_i \in [u_i,o_i]$, $i = 1(1)N$. The nodes $(x_i,y_i)$ are assumed to be not collinear.

In order to construct an interpolating function of $f(x,y)$ the convex hull of the nodes $(x_i,y_i)$ is triangulated, and then in every triangle a local polynomial $p_j(x,y)$ is defined. These local polynomials are now composed to give a continuous or even smooth differentiable global spline $S(f,x,y)$ such that $S(f,x_i,y_i) \in [u_i,o_i]$, $i=1(1)N$.

In order to obtain an applicable program for general $f(x,y)$ and $(x_i,y_i)_{i=1}^{N}$ there must be usually a priori more coefficients in $S(f,x,y)$ than conditions to define $S(f,x,y)$ uniquely. Hence a global condition is added, by which a suitable functional $\hat{N}f$ has to be minimized.

Finally the function $S(f,x,y)$ is represented and plotted by a set of curves of constant height.
The program is available in the Computer Center of the Technical University Aachen. Interested people should contact the second author.

## 2. The triangulation

The first problem is to obtain a suitable triangulation of the convex hull $\Omega$ of the nodes $(x_i,y_i),\dots,(x_N,y_N)$. There is first a theoretical point of view concerning the questions of unicity and a smallest number of triangles. Unicity is of course not given since the most trivial counterexample is the quadrilateral which is triangulated differently by the two diagonals. The smallest number of triangles depends on the distribution of the nodes. In Fig. 1 we have two examples with 11 nodes, but the number of of triangles is 9 in the first case, and 16 in the second case. If the boundary triangles are omitted (leaving the convexity!) then we have still 12 triangles.

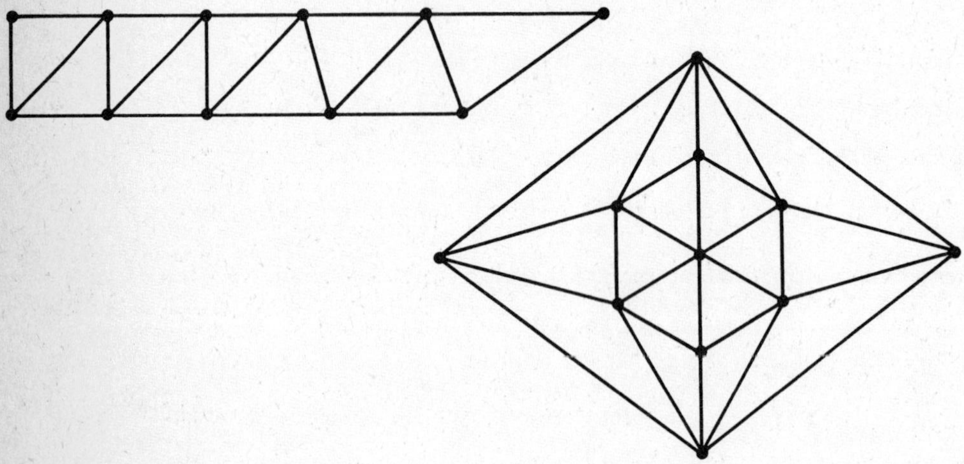

Fig. 1

The second point of view concerning the triangulation is a more practical one. It should be avoided (also with respect to error estimates as in FEM techniques) that very thin triangles similar to needles occur.

The triangulation is generated in two steps. At the beginning an arbitrary triangulation is generated, and this is improved several times. The initial triangulation is a more technical problem while the process of improving this initial triangulation is mathematically more interesting. In order to facilitate the computations we refer always to the standard triangle $D_0$ with nodes $(0,0)$ , $(1,0)$ , and $(0,1)$ . As well known (e.g. from FEM-techniques) in Fig. 2 the right one of the two triangulations is to be preferred.

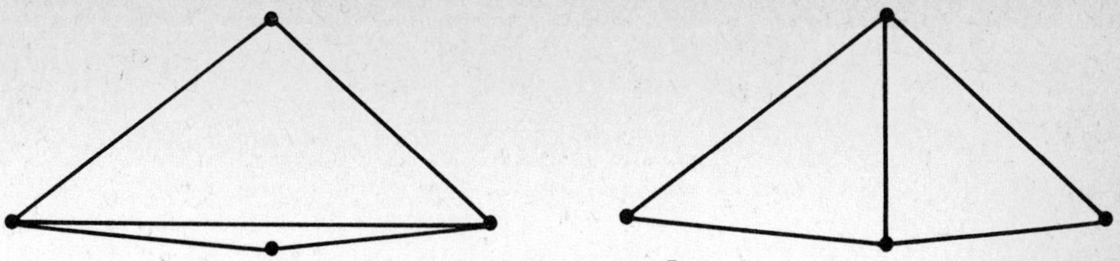

Fig. 2

So we improve the triangulation in such a way that

- the smallest angle $\varphi$ (or $\sin\varphi$ resp.) of the set of triangles becomes as large as possible.

Alternatively we consider the area $C_i$ of the osculating interior circle of a triangle with area $\Delta_i$, and calculate the ratio $R_i := C_i / \Delta_i$. Then we obtain an equivalent but in view of programming more suitable criterion

- the minimum $R$ of the ratios $R_i$ of the set of triangles becomes as large as possible.

For details of the program see [2] .

3. The interpolation

Once the triangulation is performed we construct a polynomial of a given degree on each triangle. These are then put together to a spline $S(f,x,y)$ such that the resulting "surface" of the spline $S(f,x,y)$ passes through the given intervals $[u_i,o_i]$, and in addition is as smooth as possible. So we have a certain number of polynomial coefficients in each triangle which are related by transient conditions at the boundaries of adjacent triangles, and by interpolation conditions. We cannot expect that the number of coefficients equals the number of conditions. Hence the only way to circumvent this is to have more polynomial coefficients than necessary. To this end we use a result of Zenisek [3] : The wanted spline belongs to the class $C^m(\Omega)$ for any arbitrary triangulation if the partial derivatives of degree $\leq 2m$ are used at each node. In this case the polynomials have degree at least $4m+1$. Hence for everyone of the $N$ nodes we need at least

$$P := (2m+1)\,(m+1)$$

parameters in order to determine the

$$K := (2m+1)\,(4m+3)$$

coefficients of the M polynomials in $\Omega$ having degree

$$G := 4m+1 \quad .$$

The practically most interesting case is $m = 1$ .

We now choose a suitable functional which is minimised in order to determine those coefficients which are not yet determined by corresponding conditions. This functional is

$$\hat{N}f := \sqrt{\sum_{j=1}^{M} \sum_{|\underline{s}|=o}^{m+1} \hat{a}_{j\underline{s}} \int_{D_j} [D^{\underline{s}} f(x,y)]^2 \, dx \, dy}$$

where M is the number of triangles, and $\underline{s}$ is a multiindex. The coefficients $\hat{a}_{j\underline{s}}$ allow a local weigthing of certain derivatives of f . $\hat{N}f$ gives a seminorm of f , and if $\hat{a}_{j\underline{s}} \neq 0$ it is even a norm. Hence we require that

$$\hat{N}f \overset{!}{=} \text{Min}$$

for the set of interpolating splines still containing a lot of undetermined parameters. This leads to a semidefinite quadratic form satisfaying

$$(\hat{N}p)^2 = \underline{c}^T \, \underline{\underline{M}} \, \underline{c}$$

with a Mk x Mk-matrix $\underline{\underline{M}}$ , and the vector $\underline{c}$ of coefficients of the polynomials with restrictions $u_i \leq S(f,x_i,y_i) \leq o_i$ , $i=1(1)N$ . Now it is heuristically clear how to proceed, however the detailed discussion, and the proofs are far beyond the scope of this paper.

The fundamentals for the one-dimensional case are to be found in [1] . The extension of these results to the two-dimensional case is theoretically straight-forward but it is a very tedious and complex task to perform this extension.

The program [2] performs the minimisation of $\hat{N}$ subject to the restrictions

$u_i \le S(f,x_i,y_i) \le o_i$ by use of the steepest-descent method.

## 4. Representation of the results

Once the interpolating spline $S(f,x,y)$ passing through the intervals $[u_i,o_i]$, $i=1(1)N$, is computed, we must look for a suitable way to organize the output. Usually the user is not interested in the list of coefficients of the polynomials which determine the spline. He usually likes to see an global representation of $S(f,x,y)$ which gives the really needed information. This is in most cases attained by a plotted picture of $S(f,x,y)$ such as a 3-D-plot. However, in view of measuring a representation by lines of equal height is often preferred. This is also done in [2] . The main tools are successive refinement of the triangles, and determination of zeros on their boundaries.

## 5. Numerical examples

Finally we present two examples. The first one is based on the function plotted in Fig. 3 . The position of the local maximum is at the point (0.25, 0.5) . Fig. 4 shows the triangulation with 15 nodes. The derivatives of equal order in the functional $\hat{N}f$ are weighted uniquely so that the $\hat{a}_{j\underline{s}} = \hat{a}_j$ does no longer depend on $\underline{s}$ . For $m=1$ we choose $\hat{a}_0 = \hat{a}_1 = 0$ , $\hat{a}_2 = 1$ , and obtain the results of Fig. 5 , where $|o_i - u_i| \le 0.2$ .

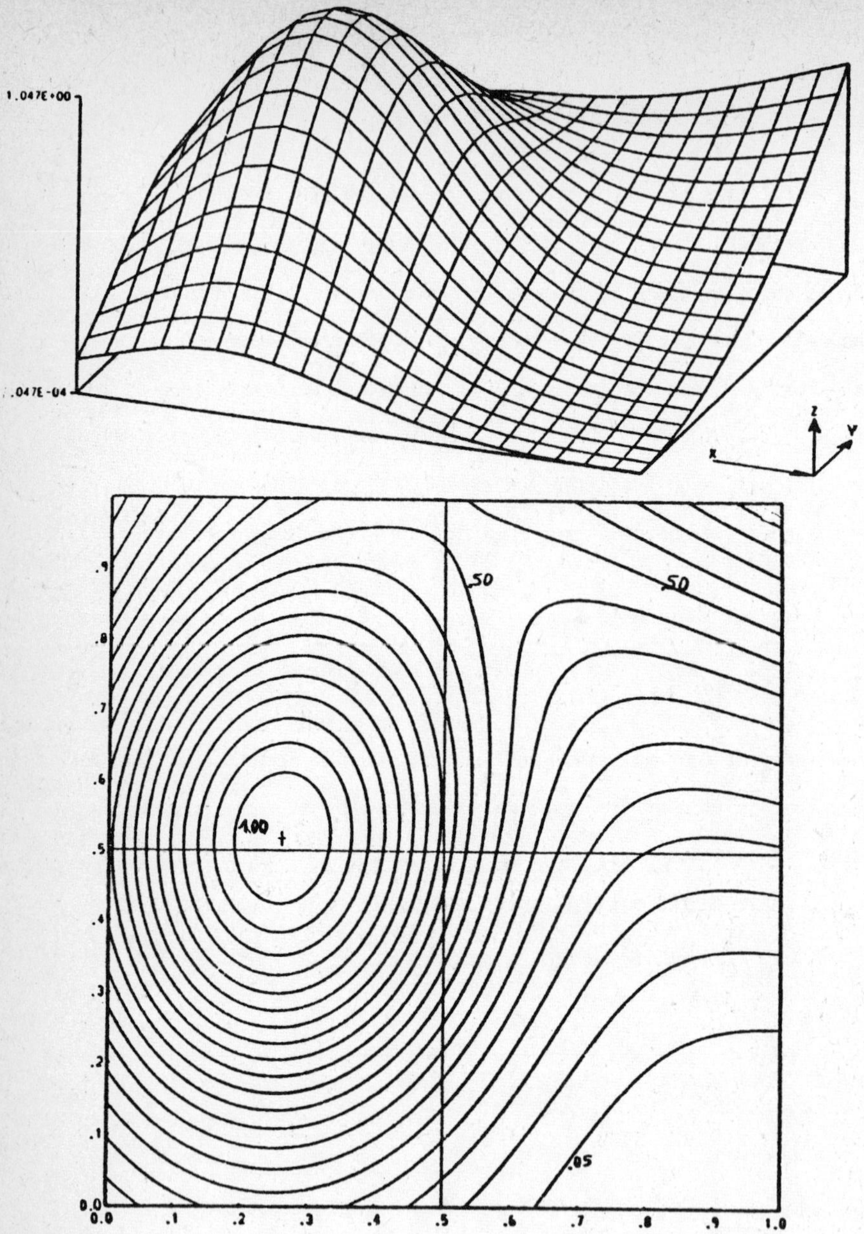


Fig. 3

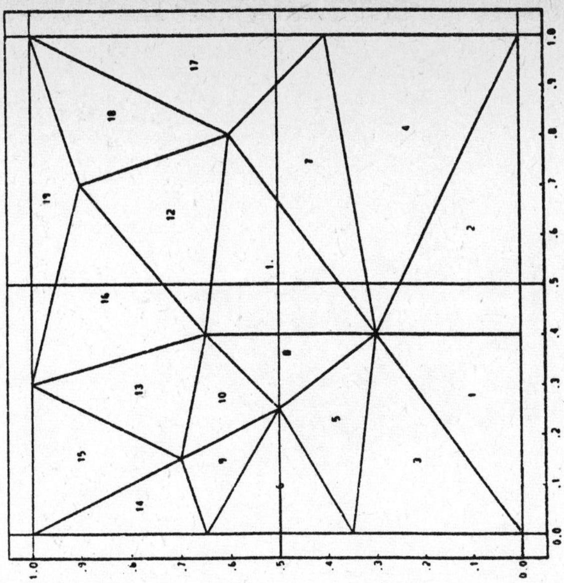
0.0
.1
.2
.3
.4
.5
.6
.7
.8
.9
1.0
1
2
3
4
5
6
7
8
9
10
11
12
13
14
15
16
17
18
19

Fig. 4

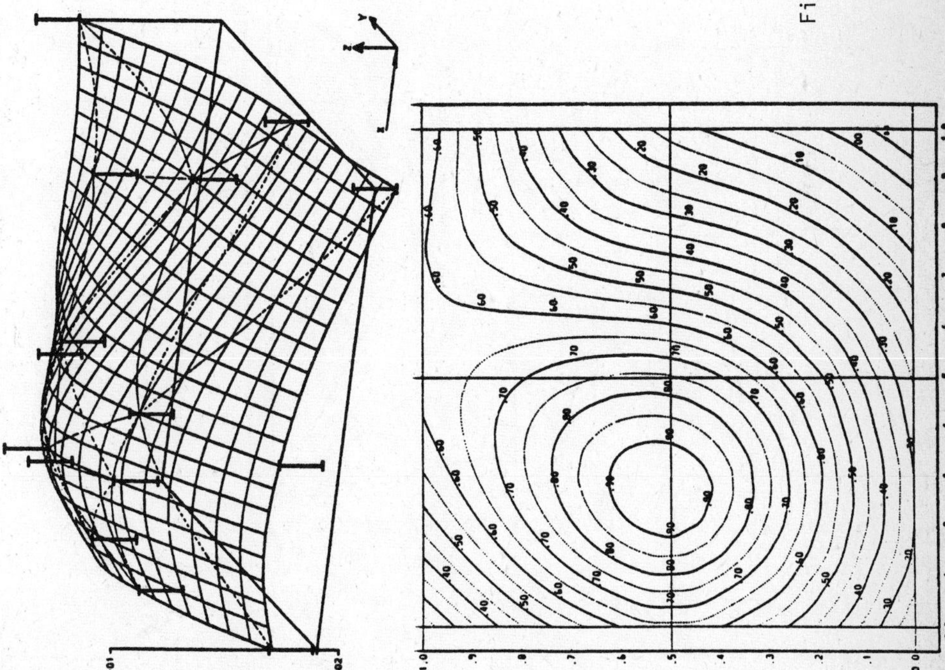
9.469E-01
0.
-9.920E-02
X
Y
Z
0.0
.1
.2
.3
.4
.5
.6
.7
.8
.9
1.0

Fig. 5

The second example concerns a peaked function. This is approximately plotted in Fig. 6 using the program with $\hat{a}_o = \hat{a}_2 = 0$ , $\hat{a}_1 = 1$ , and $u_i = o_i$ , $i=1(1)N$ 15 .

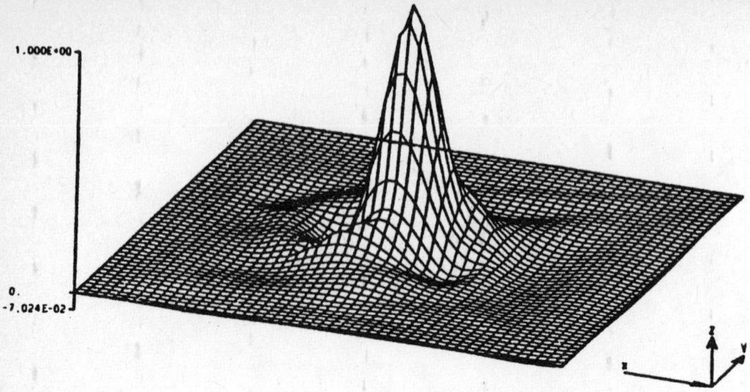


Fig. 6

The triangulation is as in Fig. 7 .

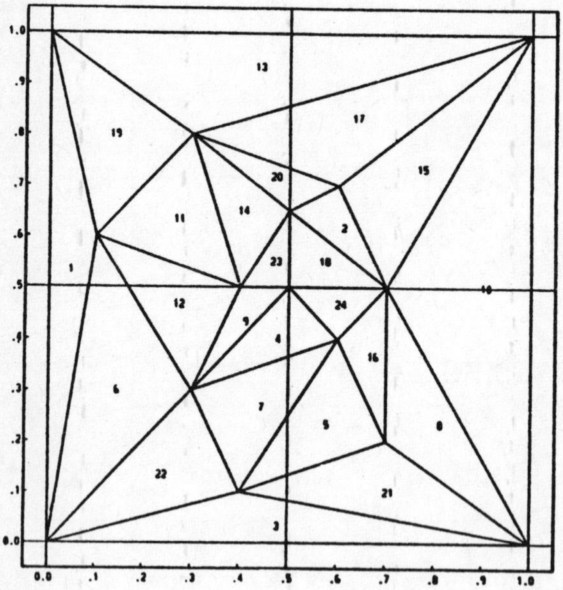


Fig. 7

In Fig. 8 and Fig. 9 we have the result for $\hat{a}_0 = \hat{a}_2 = 0$ , $\hat{a}_1 = 1$ and $|o_i - u_i| < 0.2$ .

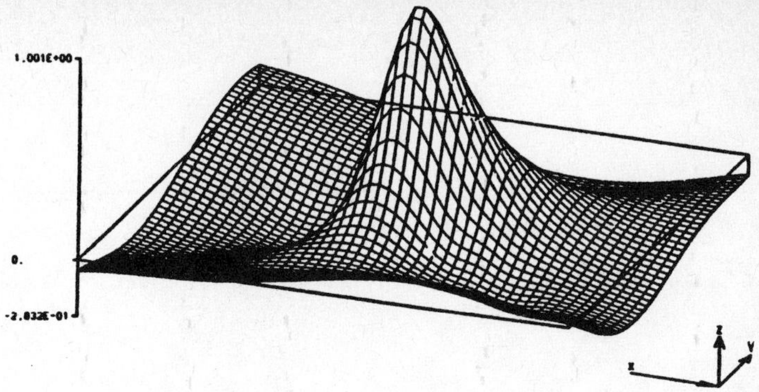


Fig. 8

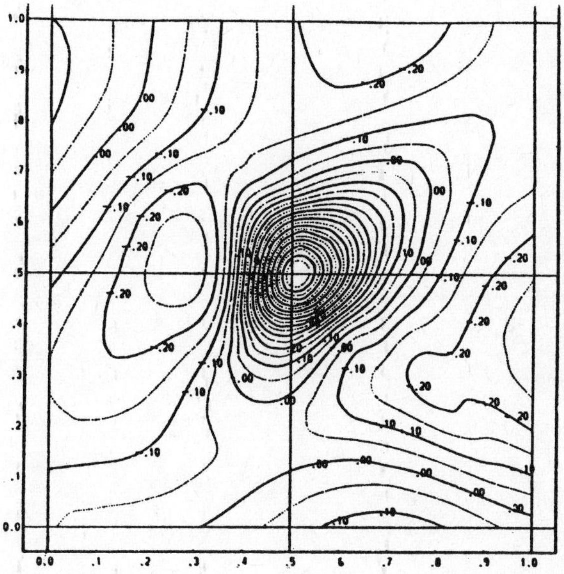


Fig. 9

We can say that this is a very "stiff" problem. The errors especially near the boundaries are not so small as one would like to see. In this case we get better results if the triangulation is refined (which causes a big additional amount of CPU-time) or by choosing a more appropriate (exponential or rational) spline.

References

[1] Groß, S. Problemorientierbare Interpolationsmethoden bei interwallwertigen Daten.
Thesis, Aachen 1980

[2] an May, D. Programm zur Erstellung eines Höhenlinienbildes einer Funktion, die durch einzelne, ungenaue Funktionswerte gegeben ist.
Computing Center of the Technical University Aachen, Seffenter Weg 23, 5100 Aachen

[3] Zenisek, A. A general theorem on triangular Finite $C^{(m)}$-Elements.
RAIRO 8 (1974), 119 - 127

## Acceptable Solutions of Linear Interval Integral Equations

Herbert Fischer
Institut für Angewandte Mathematik und Statistik
Technische Universität München
Munich, Germany

### I. Acceptable Solutions

Let us consider an integral equation of the form

$$\int_0^1 k(s,t)x(t)\,dt = y(s) \tag{1}$$

where $x \in X$, $y \in Y$, $k \in K$. We assume that $X$ and $Y$ are linear spaces of real functions on $[0,1]$ and $K$ is a suitable collection of real functions on $[0,1]^2$.

Using the linear integral operator

$$A\colon X \to Y,\ Au = v,\quad v(s) := \int_0^1 k(s,t)u(t)\,dt$$

we can rewrite the equation (1) in the form

$$Ax = y\ . \tag{2}$$

Suppose we have calculated an $\overline{x} \in X$ with $A\overline{x} \approx y$, that is, $\overline{x}$ is an approximate solution of the given problem. Now one often says:

$\overline{x}$ is <u>acceptable</u> iff $\overline{x}$ is an exact solution of a slightly disturbed problem.

This notion is common, but it is meaningless as long as "slightly disturbed" is not defined. To be precise, we have to use some kind of tolerance regions, for instance intervals. We order the function spaces $X$ and $Y$ pointwise. The set $L(X,Y)$ of all linear operators $X \to Y$ can be ordered by

$$S \le T :\iff Sx \le Tx \quad \text{for all } x \in X,\ x \ge 0\ .$$

Let LI be the set of all linear integral operators $X \to Y$. Of course the structure of LI depends on which linear operators are "integral". We defer the definition of "integral" until later. Nevertheless the order in $L(X,Y)$ induces an order in the subset LI. The order relations in X, Y, $L(X,Y)$ and LI enable us to use intervals. If H is an ordered set and $a,b \in H$ we define

$$[a,b]_H := \{h \mid h \in H,\ a \leq h \leq b\}.$$

We may drop the subscript of an interval, if the underlying ordered set is evident from the context.

For $\alpha \in LI$ with $\alpha \geq 0$ and $\eta \in Y$ with $\eta \geq 0$ it makes sense to define

$\overline{x} \in X$ is an acceptable approximate solution of problem (2) with respect to the tolerances $\alpha$ and $\eta$ iff

$$\exists\ \overline{A} \in [A-\alpha, A+\alpha]_{LI},\ \overline{y} \in [y-\eta, y+\eta]:\ \overline{A}\overline{x} = \overline{y}\ . \tag{3}$$

A quite different way to handle tolerances in problem (2) is the Interval Analysis approach.

For $\alpha \in LI$ with $\alpha \geq 0$ and $\eta \in Y$ with $\eta \geq 0$ we consider the following collection of equations:

$$\{\overline{A}x = \overline{y} \mid \overline{A} \in [A-\alpha, A+\alpha]_{LI},\ \overline{y} \in [y-\eta, y+\eta]\}\ .$$

This collection may be written symbolically as "linear interval integral equation" in the form

$$[A-\alpha, A+\alpha]_{LI}\, x = [y-\eta, y+\eta]\ . \tag{4}$$

An $\overline{x} \in X$ satisfying all equations of (4) does not exist in general. It suffices if $\overline{x} \in X$ satisfies at least one equation of (4). So we define

$\overline{x} \in X$ is an acceptable solution of problem (4) iff

$$\exists\ \overline{A} \in [A-\alpha, A+\alpha]_{LI},\ \overline{y} \in [y-\eta, y+\eta]:\ \overline{A}\overline{x} = \overline{y}\ . \tag{5}$$

Notice that the formulas (3) and (5) are identical. For the special case of equations in $L^p$ - spaces we shall present a means to check whether or not an $\overline{x} \in X$ is acceptable.

II. Lemma

In this section we start anew. The notation here does not depend on section I.

Let X be a Riesz - space and let Y be a Dedekind - complete Riesz - space. Further let L(X,Y) denote the set of all linear operators $X \to Y$, ordered by

$$S \le T \quad :\iff \quad Sx \le Tx \quad \text{for all } x \in X^+ . \tag{6}$$

The subset $LB \subseteq L(X,Y)$ of all linear order - bounded operators $X \to Y$ is a Dedekind - complete Riesz - space. For the definitions and properties of Riesz - spaces we refer to Luxemburg/Zaanen [2] and Vulikh [6]. The following lemma is a generalization of the Oettli/Prager - Theorem [3].

Lemma

For $\overline{x} \in X$, $y \in Y$, $\eta \in Y^+$, $A \in LB$, $\alpha \in LB^+$ the following assertions are equivalent

(a) $\exists\, \overline{A} \in [A-\alpha, A+\alpha]_{LB}$, $\overline{y} \in [y-\eta, y+\eta]$: $\overline{A}\overline{x} = \overline{y}$

(b) $|A\overline{x} - y| \le \alpha|\overline{x}| + \eta$

Proof.

(a) ⇒ (b): This implication is simple to prove.

From (a) we know $|\overline{A} - A| \le \alpha$ and $|\overline{y} - y| \le \eta$. So we get

$$|A\overline{x} - y| = |A\overline{x} - \overline{A}\overline{x} + \overline{y} - y| \le |(A - \overline{A})\overline{x}| + |\overline{y} - y| \le |A - \overline{A}| \cdot |\overline{x}| + \eta \le \alpha|\overline{x}| + \eta$$

(b) ⇒ (a): The Oettli/Prager proof can not be used.

The case $\overline{x} = 0$ is trivial: put $\overline{A} := A$ and $\overline{y} := 0$. In the following we assume $\overline{x} \neq 0$.

(b) supplies an interval for $A\overline{x}$,

$$A\overline{x} \in [y - \alpha|\overline{x}| - \eta, y + \alpha|\overline{x}| + \eta] = [-\alpha|\overline{x}|, \alpha|\overline{x}|] + [y - \eta, y + \eta] . \tag{7}$$

Here we use a well - known formula for adding intervals in a Riesz - space, see Schaefer [4] p. 207. From (7) we conclude

$$\exists\ \bar{z} \in [-\alpha|\bar{x}|,\alpha|\bar{x}|],\quad \bar{y} \in [y-\eta,y+\eta]:\ A\bar{x} = \bar{z} + \bar{y}\ . \tag{8}$$

Setting $B: \mathbb{R}\bar{x} \to Y$ linear with $B\bar{x} := -\bar{z}$, we obtain

$$A\bar{x} + B\bar{x} = \bar{y}\ . \tag{9}$$

The operator B is only defined on the one-dimensional linear subspace $\mathbb{R}\bar{x}$ of X.

Now we need a suitable extension of B.

The mapping $s: X \to Y$ with $s(x) := \alpha|x|$ is sublinear and for $x = \xi\bar{x} \in \mathbb{R}\bar{x}$ we get

$$Bx \le |Bx| = |B(\xi\bar{x})| = |\xi|\cdot|B\bar{x}| = |\xi|\cdot|\bar{z}| \le |\xi|\cdot\alpha|\bar{x}| = \alpha|\xi\bar{x}| = \alpha|x| = s(x).$$

Here we use $B\bar{x} = -\bar{z}$ (according to the definition of B) and $|z| \le \alpha|\bar{x}|$, which follows from (8). Thus we have

$B: \mathbb{R}\bar{x} \to Y$ linear with $Bx \le s(x)$ for all $x \in \mathbb{R}\bar{x}$.

Using the Hahn/Banach extension theorem for linear operators into Dedekind-complete Riesz-spaces, see Jameson [1] p. 64, we extend B to

$\bar{B}: X \to Y$ linear with $\bar{B}x \le s(x)$ for all $x \in X$.

We deduce some properties of the operator $\bar{B}$.

$\bar{B}x \le s(x) = \alpha|x| = \alpha x$ for all $x \in X^+$, so $\bar{B} \le \alpha$.

$(-\bar{B})x = \bar{B}(-x) \le s(-x) = \alpha|-x| = \alpha|x| = \alpha x$ for all $x \in X^+$, so $-B \le \alpha$.

Hence $|\bar{B}| = \sup\{\bar{B},-\bar{B}\} \le \alpha$ and $\bar{B} \in LB$.

Now we rewrite (9) using the extension $\bar{B}$ of B,

$$A\bar{x} + \overline{B\bar{x}} = \bar{y}$$

$$(A+\bar{B})\bar{x} = \bar{y}$$

If we set $\bar{A} := A + \bar{B}$, we obtain

$$\overline{A\bar{x}} = \bar{y}$$

and in addition $|\bar{A} - A| = |\bar{B}| \le \alpha$, $\bar{A} \in [A-\alpha,A+\alpha]$. QED

## III. Special Case

Let M denote the Riesz - space of all real Lebesgue - measurable functions on [0,1] with the usual identification of functions equal a.e. The subspaces $L^p \subseteq M$, $1 \le p < \infty$, of p-th power Lebesgue - integrable functions are Dedekind - complete Riesz - spaces and order - ideals in M. In the sequel let $1 \le p,q < \infty$. The set $L(L^p,L^q)$ of all linear operators $L^p \to L^q$ is ordered as defined in section II.

As far as linear <u>integral</u> operators $L^p \to L^q$ are concerned we follow Schep [5].

<u>Definition.</u> The linear operator $U: L^p \to L^q$ is called <u>integral</u> if there exists a Lebesgue - measurable function k on $[0,1]^2$ such that

a) $(Ux)(s) = \int_0^1 k(s,t)x(t)\,dt$ a.e. on [0,1] for all $x \in L^p$,

b) $\int_0^1 |k(s,t)x(t)|\,dt$ represents an element of $L^q$ for all $x \in L^p$.

The subset $LI \subseteq L(L^p,L^q)$ of all linear integral operators $L^p \to L^q$ carries the order inherited from $L(L^p,L^q)$.

Now we consider the linear interval integral equation

$$[A-\alpha,A+\alpha]_{LI}x = [y-\eta,y+\eta]_{L^q}, \qquad (10)$$

where $A \in LI$, $\alpha \in LI^+$, $y \in L^q$, $\eta \in L^{q+}$ are given.

In the following theorem we show that an $\bar{x} \in L^p$ is an acceptable solution of the problem (10) iff $|A\bar{x} - y| \le \alpha|\bar{x}| + \eta$.

<u>Theorem.</u>

For $\bar{x} \in L^p$, $y \in L^q$, $\eta \in L^{q+}$, $A \in LI$, $\alpha \in LI^+$

the following assertions are equivalent

(a) $\exists\, \bar{A} \in [A-\alpha,A+\alpha]_{LI},\ \bar{y} \in [y-\eta,y+\eta]_{L^q}: \bar{A}\bar{x} = \bar{y}$

(b) $|A\bar{x} - y| \le \alpha|\bar{x}| + \eta$ .

Proof. We use our Lemma and a result of Schep [5]. Schep showed that LI is a band in the Riesz-space LB of all linear order-bounded operators $L^p \to L^q$.

(a) ⇒ (b): Because of LI $\subseteq$ LB the operator $\bar{A}$ is order-bounded. Then proceed as in (a) ⇒ (b) of the Lemma.

(b) ⇒ (a): Because of LI $\subseteq$ LB, we may apply the Lemma. This yields

$$\exists\ \bar{A} \in [A-\alpha, A+\alpha]_{LB},\ \ \bar{y} \in [y-\eta, y+\eta]_{L^q} : \bar{A}x = \bar{y} \tag{11}$$

It is clear that $[A-\alpha, A+\alpha]_{LI} \subseteq [A-\alpha, A+\alpha]_{LB}$.

This inclusion is in fact an equality:

An arbitrary element C in $[A-\alpha, A+\alpha]_{LB}$ can be written in the form C = A + B with B $\in$ LB, $|B| \leq \alpha$. LI as a band in LB is in particular an order-ideal in LB. So $|B| \leq \alpha$ implies B $\in$ LI. From C = A + B $\in$ LI and $|B| \leq \alpha$ we get C $\in [A-\alpha, A+\alpha]_{LI}$.

Hence $[A-\alpha, A+\alpha]_{LI} = [A-\alpha, A+\alpha]_{LB}$.

With this information at hand we rewrite (11):

$$\exists\ \bar{A} \in [A-\alpha, A+\alpha]_{LI},\ \ \bar{y} \in [y-\eta, y+\eta]_{L^q} : \bar{A}x = \bar{y}\ .$$

QED

References


[1] Jameson, G., Ordered Linear Spaces.
Lecture Notes in Mathematics, Vol. 141, Springer 1970.

[2] Luxemburg/Zaanen, Riesz Spaces I.
North Holland Publ. Comp. 1971.

[3] Oettli/Prager, Compatibility of Approximate Solution of Linear Equations with Given Error Bounds for Coefficients and Right-Hand Sides.
Numer. Math. 6, 1964, 405 - 409.

[4] Schaefer, H.H., Topological Vector Spaces.
3. Printing, Springer 1971.

[5] Schep, A.R., Kernel Operators.
Indagationes Mathematicae 41, 1979, 39 - 53.

[6] Vulikh, B.Z., Introduction to the Theory of Partially Ordered Spaces.
Wolters-Noordhoff Sc. Publ. 1967.

# MAXIMIZATION OF MULTIVARIABLE FUNCTIONS USING INTERVAL ANALYSIS

Yasuo Fujii
Educational Center for Information Processing
Kyoto University

Kozo Ichida
Faculty of Business Administration
Kyoto Sangyo University

Masahiro Ozasa
Department of Electrical Engineering
Ritsumeikan University

## I. INTRODUCTION

A number of techniques have been proposed for nonlinear optimization problems. Some of them are conjugate gradient method, simplex method, variable metric method and random search method [1].

However, if the objective function is multimodal, we have few methods for finding the global maximum or minimum [2].

Interval analysis is very effective for this global optimization problems[3],[4],[5]. The simple way to apply this method is to divide the original domain into subregions, and to delete subregions that can not have the global maximum [6]. This algorithm is not very efficient since a great many of subregions remain without being discarded.

In this paper we describe an interval method to compute the global maximum value of the multivariable function over the hyperrectangle.

The interval Newton method is used for finding the stationary points in the domain and on the boundary. On the boundary one or more variables are fixed as constants, so that the dimension of the Hessian matrix decreases.

The constrained optimization with equality or inequality condition can be solved by the Lagrange multiplier method.

Our interval arithmetic system is written with FORTRAN 77 and the assembly language. The upper and the lower bounds of the number can be calculated at an arbitrary digit.

## II. UNCONSTRAINED MAXIMIZATION

### II-1. One-Dimensional Case

We consider how to compute the greatest value of the function f(x) on a closed interval [a, b], where f is supposed to belong to $C^2$. The global maximum is obtained by computing the maximum of relative maxima in (a, b) and function values at the two end points.

The interval Newton algorithm is used to obtain the stationary values of f(x) by solving the equation f'(x)=0 [7]. It is given as

$$(1)\qquad \left.\begin{aligned} N(X_p) &= m(X_p) - \frac{F'(m(X_p))}{F''(X_p)}, \\ X_{p+1} &= X_p \cap N(X_p), \end{aligned}\right\} \quad (p=0,1,\ldots)$$

where capital letters X,F',F" denote interval extensions of x,f',f" respectively and $m(X_p)$ is the midpoint of $X_p$. The maximization algorithm for one-variable function is as follows.

Step 1: Compute $f^* = \max[f(a),f(b)]$ and the point(s) $x^*$ at which $f(x^*) = f^*$.

Step 2: Divide the original interval A=[a, b] into two subintervals at the midpoint m=(a+b)/2.

Step 3: If the widths of the undiscarded interval(s) are sufficiently small, then stop. Otherwise pick out an interval (let it be $A_i$), and calculate $F'(A_i)$ and $F''(A_i)$.

Step 4: If $F'(A_i) \not\ni 0$, then there is no stationary point in $A_i$. So delete $A_i$ and go to Step 3. Otherwise go to Step 5.

Step 5: If $F''(A_i) > 0$, then the stationary value is a relative minimum and $A_i$ can be discarded. If $F''(A_i) < 0$, compute this relative maximum by use of the interval Newton method. Replace f* and x* if the obtained relative maximum is greater than f*. If $F''(A_i) \ni 0$, divide $A_i$ into two at its midpoint and go to Step 3.

## II-2. Multi-Dimensional Case

Compute the maximum of multimodal function $f(x_1, x_2, \ldots, x_k)$ in $C^2$ over a k-dimensional box $\mathbf{A} = [a_1, b_1] \times \ldots \times [a_k, b_k]$. To seek the maximum of f, it is necessary to compute the values of relative maxima in $\mathbf{A}$ and maximum values on the boundary of $\mathbf{A}$. As compared with the one-dimensional case, we have the following difficulty.

[i] In the one-dimensional case we can distinguish relative maximum from relative minimum. In the multi-dimensional case the sign of the Hessian can not distinguish relative maximum from minimum or saddle point.

[ii] In the one-dimensional case the boundary consists of only two end points. In the multi-dimensional case the boundary becomes the lower-dimensional region on which the maximum must be sought.

As concerns [i] we calculate each stationary value, and replace the maximum value so far obtained with this value if it is larger.

Concerning [ii] we apply the interval Newton method in various dimensions (from 1 to k-1) to seek the stationary value on the boundary. For example if the original domain is four-dimensional, its boundary becomes three-dimensional on which three-dimensional Newton method is applied. Moreover the boundary of this three-dimensional region becomes two-dimensional, and so on. It should be noted that even if a region is deleted since no global maximum exists in it, we must save its boundary if it has a common boundary with the original domain.

The multi-dimensional interval Newton method corresponds to equation (1) is given as

$$\text{(2)} \qquad \left.\begin{aligned} \mathbf{N}(X_n) &= \mathbf{m}(X_n) - J^{-1}(\mathbf{X}_n)\nabla F(m(\mathbf{X}_n)), \\ \mathbf{X}_{n+1} &= \mathbf{X}_n \cap \mathbf{N}(\mathbf{X}_n), \end{aligned}\right\} (n=0,1,\ldots)$$

$$\text{(3)} \qquad \nabla F = \left(\frac{\partial F}{\partial X_1}, \frac{\partial F}{\partial X_2}, \ldots, \frac{\partial F}{\partial X_k}\right)^T ,$$

$$(4) \qquad J = \begin{bmatrix} \frac{\partial^2 F}{\partial X_1^2} & \frac{\partial^2 F}{\partial X_1 \partial X_2} & \cdots\cdots & \frac{\partial^2 F}{\partial X_1 \partial X_k} \\ \frac{\partial^2 F}{\partial X_2 \partial X_1} & \frac{\partial^2 F}{\partial X_2^2} & \cdots\cdots & \frac{\partial^2 F}{\partial X_2 \partial X_k} \\ \vdots & \vdots & & \vdots \\ \frac{\partial^2 F}{\partial X_k \partial X_1} & \frac{\partial^2 F}{\partial X_k \partial X_2} & \cdots\cdots & \frac{\partial^2 F}{\partial X_k^2} \end{bmatrix} .$$

To get the Newton sequence (2), we set

$$(5) \qquad H_n = -J^{-1} \nabla F$$

and solve the linear equation

$$(6) \qquad J H_n = - \nabla F$$

with Gauss' elimination method.

On the boundary $x_1 = a_1$ (6) becomes the following (k-1)-dimensional equation since the elements differentiated with respect to $X_1$ vanish.

$$(7) \qquad \begin{bmatrix} 0 & 0 & \cdots\cdots & 0 \\ 0 & \frac{\partial^2 F}{\partial X_2^2} & \cdots\cdots & \frac{\partial^2 F}{\partial X_2 \partial X_k} \\ \vdots & \vdots & & \vdots \\ 0 & \frac{\partial^2 F}{\partial X_k \partial X_2} & \cdots\cdots & \frac{\partial^2 F}{\partial X_k^2} \end{bmatrix} \begin{bmatrix} 0 \\ H_2 \\ \vdots \\ H_k \end{bmatrix} = - \begin{bmatrix} 0 \\ \frac{\partial F}{\partial X_2} \\ \vdots \\ \frac{\partial F}{\partial X_k} \end{bmatrix} .$$

The maximization algorithm for multi-variable function is as follows :

<u>Step 1</u>: Let f* be the maximum value of f at the $2^k$ vertices.

<u>Step 2</u>: Divide **A** at the midpoint of the maximum side to generate two

regions $A_1$ and $A_2$.

<u>Step 3</u>: If the widths of the undiscarded regions are sufficiently small, then stop. Otherwise take out a region $A_i$, and evaluate $\nabla F(A_i)$, $J(A_i)$ and $D = \det(J)$.

<u>Step 4</u>: If $0 \notin \nabla F(A_i)$, there is no stationary point in $A_i$. So discard $A_i$ and go to Step 3. However, if $A_i$ has a common boundary with $A$, then this common boundary should be saved as a (degenerate) region.

If $0 \in \nabla F(A_i)$, go to Step 5.

<u>Step 5</u>: If $0 \in D$, divide $A_i$ at the midpoint of its maximum side and go to Step 3. Otherwise apply the interval Newton method to find the stationary value. If the obtained value is greater than f*, f* is replaced with it. Then go to Step 3. If $A_i$ has a common boundary with the original domain $A$, the common boundary should be saved whether $A_i$ has a stationary point or not.

## III. CONSTRAINED MAXIMIZATION

We consider to find the value of $\mathbf{x}$ that maximizes

$$f(\mathbf{x}) = f(x_1, x_2, \ldots, x_k) \tag{8}$$

in the bounded region $[a_1, b_1] \times \ldots \times [a_k, b_k]$ subject to the r equality constraints

$$g_j(\mathbf{x}) = 0 \quad (j=1, \ldots, r). \tag{9}$$

To solve this problem we use the Lagrange function

$$L(\mathbf{x}, \mathbf{p}) = f(\mathbf{x}) + \sum_{j=1}^{r} p_j g_j(\mathbf{x}), \tag{10}$$

where $\mathbf{p} = (p_1, \ldots, p_r)^T$ is a Lagrange-multiplier [8]. The necessary condition at a maximum of the function is

$$\frac{\partial L}{\partial x_i} = \frac{\partial f}{\partial x_i} + \sum_{j=1}^{r} p_j \frac{\partial g_j(\mathbf{x})}{\partial x_i} = 0 \quad (i=1,2,\ldots,k), \tag{11}$$

(12) $$\frac{\partial L}{\partial p_j} = g_j(\mathbf{x}) = 0 \quad (j=1,2,\ldots,r).$$

We can apply the interval Newton method in section II-2 to solve these nonlinear simultaneous equations. The constrained global maximum is obtained among the stationaly values of L.

Inequality constraints can be converted to the equality constraints by introducing an extra slack variable. The expression

(13) $$g_j(\mathbf{x}) \leqq 0 \quad (j=1,\ldots,r)$$

can be written as

(14) $$g_j(\mathbf{x}) + x_{k+j}^2 = 0 \quad (j=1,\ldots,r).$$

Then Lagrange function L is written as

(15) $$L(\mathbf{x},\mathbf{p}) = f(\mathbf{x}) + \sum_{j=1}^{r} p_j(g_j + x_{k+j}^2).$$

The stationary points can be computed by the interval Newton method as before. The case that both equality and inequality constraints are contained can be treated similarly.

## IV. NUMERICAL EXAMPLES

Several numerical examples have been computed by the method described above.

The calculations were done with HITAC M200H (corresponds to IBM 3083JX) of the Educational Center for Information Processing of Kyoto University.

Example 1: Three Hump Camel-Back Function.

(16) $$f(x_1,x_2)=-2x_1^2+1.05x_1^4 - \frac{1}{6}x_1^6 - x_1x_2 - x_2^2.$$

This function is known to have three maxima and two saddle points in the domain $-5 \leq x_1 \leq 5$, $-4 \leq x_2 \leq 4$.

The computed result is:

$X_1$=[-0.49630 83675 31816 60D-22, 0.49630 83675 31816 60D-22 ],
$X_2$=[-0.70409 33880 50906 57D-22, 0.79409 33880 50906 57D-22 ],
F(max)=[-0.15173 43493 40171 69D-43, 0.19705 75965 45677 71D-45 ].

Example 2: Six Hump Camel-Back Function.

This function is known to have six maxima, two minima and seven saddle points in the domain $-5 \leq x_1 \leq 5$, $-4 \leq x_2 \leq 4$.

$$f(x_1,x_2)=-4x_1^2+2.1x_1^4-\frac{1}{3}x_1^6-x_1x_2+4x_2^2-4x_2^4. \tag{17}$$

Global maximum is obtained at the following two points. They are symmetric with respect to the origin.

$X_1$=[ 0.08984 20131 00318 030, 0.08984 20131 00318 099 ],
$X_2$=[-0.71265 64030 20739 90 , -0.71265 64030 20739 40 ],
F(max)=[ 1.03162 84534 89873 6 , 1.03162 84534 89881 2 ].

$X_1$=[-0.08984 20131 00318 099, -0.08984 20131 00318 030 ],
$X_2$=[ 0.71265 64030 20739 40 , 0.71265 64030 20739 90 ],
F(max)=[ 1.03162 84534 89873 6 , 1.03162 84534 89881 2 ].

Example 3: Five-variable Function [9].

$$f(x_1,x_2,\ldots,x_5)=f_1(x_1)f_2(x_2)f_3(x_3)f_4(x_4)f_5(x_5), \tag{18}$$

where

$$f_1(x_1)=x_1(x_1+13)(x_1-15)*0.01,$$
$$f_2(x_2)=(x_2+15)(x_2+1)(x_2-8)*0.01,$$
$$f_3(x_3)=(x_3+9)(x_3-2)(x_3-9)*0.01,$$
$$f_4(x_4)=(x_4+11)(x_4+5)(x_4-9)*0.01,$$
$$f_5(x_5)=(x_5+9)(x_5-9)(x_5-10)*0.01.$$

Case (i): $-10 \leq x_i \leq 10$, $(i=1,2,\ldots,5)$.

This function has $2^4$ maxima and $2^4$ minima in this domain.

The computed result is:

$X_1$=[ 8.75644 07330 07731 2, 8.75644 07330 07731 2 ],
$X_2$=[-9.35828 66332 94911 5, -9.35828 66332 94911 5 ],
$X_3$=[-4.57207 78818 33903 6, -4.57207 78818 33903 6 ],

$X_4$=[ 3.59212 96115 43726 2, 3.59212 96115 43726 2 ],
$X_5$=[-2.84008 63924 84045 8, -2.84008 63924 84043 4 ],
F(max)=[ 24416.03065 50573 60 , 24416.03065 50574 10 ].

Case (ii): $-10 \leq x_1 \leq 8$, $-10 \leq x_i \leq 10$, (i=2,3,4,5).
The maximum value is obtained on the boundary $x_1=8$.

$X_1$=[ 8.00000 00000 00000 0, 8.00000 00000 00000 0 ],
$X_2$=[-9.35828 66332 95584 2, -9.35828 66332 94234 0 ],
$X_3$=[-4.57207 78818 50185 3, -4.57207 78818 17701 7 ],
$X_4$=[ 3.59212 96114 94123 9, 3.59212 96115 93839 5 ],
$X_5$=[-2.84008 63924 86292 4, -2.84008 63924 81793 2 ],
F(max)=[ 24139.85650 22284 47 , 24139.85650 22284 95 ].

Case (iii): $-10 \leq x_1 \leq 8$, $-10 \leq x_i \leq 12$, (i=2,3,4), $-10 \leq x_5 \leq 10$.
The maximum value is obtained on the boundary $x_2=x_3=x_4=12$.

$X_1$=[-7.42310 73996 74397 9, -7.42310 73996 74397 9 ],
$X_2$=[12.00000 00000 00000 , 12.00000 00000 00000 ],
$X_3$=[12.00000 00000 00000 , 12.00000 00000 00000 ],
$X_4$=[12.00000 00000 00000 , 12.00000 00000 00000 ],
$X_5$=[-2.84008 63924 84044 7, -2.84008 63924 84044 7 ],
F(max)=[ 90193.85088 59564 90 , 90193.850088 59567 52 ].

Example 4: Equality constrained problem.
Maximize

$$(19) \qquad f(x_1,x_2,x_3)=-(x_1+x_2-x_3-1)^2-(x_1+x_2)^2-5x_1^2$$

subject to $2x_1+x_3=0$, $-10 \leq x_i \leq 10$, (i=1,2,3).
The computed result is:

$X_1$=[ 0.14285 71428 57138 21, 0.14285 71428 57148 04 ],
$X_2$=[ 0.21428 57142 85700 95, 0.21428 57142 85725 15 ],
$X_3$=[-0.28571 42857 14316 56, -0.28571 42857 14258 47 ],
F(max)=[-0.35714 28571 42908 72, -0.35714 28571 42803 86 ].

The Lagrange-multiplier obtained is:

P =[-0.71428 57142 85766 82, -0.71428 57142 85662 45 ].

Example 5: Inequality constrained problem.
Maximize

(20) $f(x_1,x_2)=2x_1^2-2x_1x_2+2x_2^6-6x_1$

subject to $3x_1+4x_2 \leqq 6$, $-10 \leqq x_i \leqq 10$, $(i=1,2)$.

The computed result is:

$X_1$=[ 1.45945 94594 59459 4 , 1.45945 94594 59459 4 ],

$X_2$=[ 0.40540 54054 05405 44, 0.40540 54054 05405 46 ],

F(max)=[ 5.35135 13513 51348 2 , 5.35135 13513 51354 0 ],

The values of Lagrange-multiplier p and slack valiable $x_3$ are:

$X_3$=[ 0.0 , 0.10297 26133 56609 95D-24],

P =[ 0.32432 43243 24324 40, 0.32432 43243 24324 40 ].

## V. CONCLUSION

We described an algorithm for maximizing functions by use of interval analysis. It enables us to obtain the maximum in the domain or on the boundary. Both unconstrained and constrained global maximum can be computed. So far we have calculated maxima of the functions up to five variables. If effective devices for reducing interval width of functions are developed, this method can be applied to higher-dimensional problems.

## REFERENCES

[1] Bremermann,H.:A method of unconstrained global optimization, Math. Biosci. 9,1-15(1970).

[2] Dixon,L.C.W., Szëgo,G.P.:Towords Global Optimisation 2. North-Holland/American Elsevier, Amsterdam, 1978.

[3] Hansen,E.R.:Global optimization using interval analysis —The multi-dimensional case, Numer. Math. 34,247-270(1980).

[4] Alefeld,G., Herzberger,J.:Introduction to Interval Computations. Academic Press, New York, 1983.

[5] Nickel,K.L.E.:Interval Mathematics 1980. Academic Press, New York, 1980.

[6] Ichida,K., Fujii,Y.:An interval arithmetic method for global optimization, Computing 23,85-97(1979).

[7] Moore,R.E.:Interval Analysis. Englewood Cliffs, N.J., Prentice-

Hall, 1966.
[8] Dixon,L.C.W.:Nonlinear Optimisation. The English Univ. Press, London, 1972.
[9] Tsuda,T., Sato,M.:An algorithm to locate the greatest maxima of multi-variable functions, Infomation Processing(Joho Shori) 16, 2-6(1975).

# MODAL INTERVALS : REASON AND GROUND SEMANTICS

Ernest Gardeñes
Honorino Mielgo
Albert Trepat

Facultad de Matematicas
Universidad de Barcelona
Gran Via 585 , 08007 BARCELONA / SPAIN



## 1. SEMANTIC RELATION BETWEEN INTERVALS AND REALS.

The structure STR(RE) of real numbers ( would "ideal numbers" be a better name, to emphasize their being out of reach for digital computing? ) , is the seat of geometrical intuition , which is based on the system of relations and operations that allow the construction of predicates P : RE --> SET(FALSE,TRUE) .

Otherwise, there is no computing system able to use the full STR(RE) and no finite digital set DI << RE ( << is "included in" ) closed for the whole system of exact arithmetical operations. This fact makes the structure STR(I(DI)) of digital intervals with outer rounding, into the only support for digital computing supplying the maximal approximating-information accessible to the system STR(DI) ; and STR(I(RE)) into its analytical frame.

The computing/analytic system STR(I(RE),I(DI)) displays, however , some critical problems ; let us look at three paradygmatic examples.

First : supposing F' <* I(RE) ( <* is "belongs to" ) to be the exact result of an interval computation , its outer digital approximation FO >> F' ( >> is "includes" ) keeps the validity of any predicate of the form E(f,F')P(f) when FO substitutes F' ( E(f,F') is "exists f <* F' such that" ) ; but, in order to keep the validity of predicates of the form U(f,F')P(f) ( U(f,F') is "for every f <* F' " ), an inner digital approximation FI << F' ought to be used. The problem is that, though any F' <* I(RE) bounded by the system I(DI) admits an outer rounding FO, this property does not hold for the inner rounding FI ( e.g. , the inner rounding of any x <* RE , x -<* DI , does not exist in DI ( -<* is "does not belong to" ) ).

Second : the lack of inner rounding in STR(I(RE),I(DI)) is also a drawback for the computation of the approximated interval solution of

I(RE)-systems like ( A' + X' = B' , X' + Z' = C' ) where an inner rounded X' were needed to compute an outer rounded Z'.

Third : when the solution of the equation A' + X' = B' in I(RE) exists, the relation A' + X1' = B' is equivalent to the proposition " B' is the inclusion-least interval for which U(a,A') U(x,X1') a + x <* B' holds " ; but even when the interval solution of the equation A' + X' = B' fails to exist in the I(RE) context , an interval X" <* I(RE) does exist validating the proposition " B' is the inclusion-least interval for which U(a,A') E(x,X") a + x <* B' holds" .

These three "non-sequitur" situations stated in terms of the system STR(I(DI),I(RE)) , are evidence enough to undermine the validity of this system as universal frame for the numerical-computing theory.

To find out what is missing , let us analize the relation standing among intervals , real numbers , interval predicates , and predicates about real numbers.

Two-variable predicates like ( P(x), x <* X' ) , ( (pred,pred,...) is "pred AND pred AND ... " ) , maybe would convey some P(.)-semantics from x <* RE to X' <* I(RE) , but the resulting semantics would be ambiguous for an interval argument X' because of the different truth values that P(.) could take for different points x <* X' ; moreover, this predicates would have only a designational value and would be out of question in a computational context , because of the unability to reach a general x <* X' by means of DI.

But classical interval-predicates can be obtained from real-predicates P(x) , without any reference to particular x <* RE , by means of the transformations P(x) --> E(x,X')P(x) and P(x) --> U(x,X')P(x) which transport the meanings defined by the predicates P(.) , from the domain RE to the domain I(RE).

Actually , the semantic transformation SEM : P(x) -> Q'(x,X')P(x) ( Q' <* SET(E,U) ) , brings predicates P(x) of the single real arguments x <* X' into predicates P*((X',Q')) := Q'(x,X')P(x) about the arguments X = (X',Q') <* I*(RE) , I*(RE) := CART(I(RE),SET(E,U)) ( CART is "Cartesian product" ) , which we will name "modal intervals" ( := is "defined by") .

Indeed , if for X = (X',QX) we define SET(X) := X' and MOD(X) := QX ( we will name SET(X) the set-component of X and MOD(X) its modality ), the meaning of the predicate P*(X) = Q(x,X)P(x) is fully determined by the definition : Q(x,X) :=

( IF MOD(X) = E THEN E(x,X') ,
IF MOD(X) = U THEN U(x,X') ) .

## 2. INTERVAL-SETS OF PREDICATES AND MODAL INCLUSION.

Let be PRED((X',QX)) := SET(P(.)/(Q(x,X)P(x)) the set of real predicates validated by the modal interval X = (X',QX) , and let us examine which are the conditions standing between the modal intervals A and B that correspond to the set inclusion PRED(A) << PRED(B).

LEMMA 2.1 PRED((A',E)) << PRED((B',E)) <==> A' << B'

Since A' << B' implies that ( xl <* A' , P(xl) ) ==> ( xl <* B' , P(xl) ) obviously ; and if A' -<< B' , E(a,A') a -<* B' and (x=a) <* PRED((A',E)) , but (x=a) -<* PRED((B',E)) and , therefore , PRED((A',E)) -<< PRED((B',E)).

LEMMA 2.2 PRED((A',U)) << PRED((B',U)) <==> A' >> B'

Since A'>>B' implies that U(x,A')P(x) ==> U(x,B')P(x) and if A' ->> B' , then E(b,B') b -<* A' and (x <* A') <* PRED((A',U)) but (x <* A') -<* PRED((B',U)) and therefore PRED((A',U)) -<< PRED((B',U)).

LEMMA 2.3 PRED((A',U)) << PRED((B',E)) <==> A' =* B'
( =* is "intersects" )

Since A' =* B' implies that U(x,A')P(x) ==> E(x,B')P(x) ; and if A' -=* B' then (x <* A') <* PRED((A',U)) but (x <* A') -<* PRED((B',E)) and therefore PRED((A',U)) -<< PRED((B',E)).

LEMMA 2.4 PRED((A',E)) << PRED((B',U)) <==> A' = B' = INT(a)
( INT(a) is "the point-interval with a = inf = sup " )

Since , if al <* A' then (x=al) <* PRED((A',E)) and the only possibility for the validity of U(x,B')(x=al) is that B' = INT(al) ; but in this case if a2 -= al , a2 <* A' , would exist , the predicate (x=a2) would be validated by (A',E) but not by (B',U) . The reverse implication is obvious .

DEFINITION 2.1 For A = (A',QA) , B = (B',QB) modal intervals ,
A << B := IF QA = QB = E THEN A' << B'
IF QA = QB = U THEN A' >> B'
IF ( QA = U , QB = E ) THEN A' =* B'
IF ( QA = E , QB = U ) THEN A' = B' = INT(a)

DEFINITION 2.2 For A = (A',QA) <* I*(RE) ,
INF(A) := IF QA = E THEN INF(A')
IF QA = U THEN SUP(A')
SUP(A) := IF QA = E THEN SUP(A')
IF QA = U THEN INF(A')

THEOREM 2.1 For A , B modal intervals ,
( INF(A) = INF(B) , SUP(A) = SUP(B) ) <==> A = B

DEFINITION 2.3 For a , b <* RE ,
INT(a,b) := ELEM( A / A <* I*(RE) , INF(A) = a , SUP(A) = b )
( where ELEM( A / C ) is the element named A fulfilling the condition C )

DEFINITION 2.4

Ie(RE) := SET((A',Q') / A' <* I(RE) , Q' = E )
Iu(RE) := SET((A',Q') / A' <* I(RE) , Q' = U )
Ip(RE) := SET((A',Q') / A' <* I(RE) , INF(A') = SUP(A') )

THEOREM 2.2 I*(RE) <--> SET((a,b) / a , b <* RE )
Ie(RE) = SET( A / A <* I*(RE) , INF(A) <= SUP(A) )
Iu(RE) = SET( A / A <* I*(RE) , INF(A) >= SUP(A) )
Ip(RE) = SET( A / A <* I*(RE) , INF(A) = SUP(A) )

THEOREM 2.3 A <* I*(RE) ==> PRED(A) -= VOID
( VOID is "the void set" )

Since , when A = (A',E) then ( x = INF(A) ) <* PRED(A) ,
and when A = (A',U) then ( x <* A' ) <* PRED(A) .

And the above lemmata and definitions yield easily the following theorems for A , B , ... <* I*(RE) .

THEOREM 2.4 A << B <==> ( INF(A) >= INF(B) , SUP(A) <= SUP(B) )

THEOREM 2.5 A << B <==> PRED(A) << PRED(B)

THEOREM 2.6 A = B <==> ( A << B , A >> B )
<==> PRED(A) = PRED(B)

Theorems 2.1 to 2.6 , by displaying the association (a1,a2) <--> PRED(INT(a1,a2)) , provide the lattice completion of the inclusion structure of ordinary intervals with a definitive semantical meaning , and suggest to interpret the elements of I*(RE) as acceptors/rejectors or interval-tests for the predicates about the reals , and to read "A << B" as "A is more strict than B" or "B is more tolerant than A" .

Maybe this semantics is clarified by the observation that for A <* I*(RE) , if A is a proper or exitencial modal interval ( that is A <* Ie(RE) ) then P(x) <* PRED(A) is equivalent to SET( x / P(x) ) =* SET(A) , and if B is an improper or universal modal interval ( that is B <* Iu(RE) ) then P(x) <* PRED(B) is now equivalent to SET(B) << SET( x / P(x) ) .

DEFINITION 2.5 For A <* I*(RE) ,

PROP(A) := INT( MIN(INF(A),SUP(A)) , MAX(INF(A),SUP(A)) )
IMPR(A) := INT( MAX(INF(A),SUP(A)) , MIN(INF(A),SUP(A)) )

The denomination PROP(A) comes from naming "proper intervals" the elements of Ie(RE) , or existencial intervals , because of their identification to the corresponding elements of I(RE) that arises from the equivalence in Ie(RE) of A << B and SET(A) << SET(B) . This identification keeps its force along the whole theory about I*(RE) , since the relation << in I*(RE) generates all the structure STR( I*(RE) , I*(DI) ) . Moreover , the "proper intervals" are the interval-acceptors of the "exact" real solutions that the ordinary-interval approximations are meant to bound .

## 3. DUAL SEMANTICS OF MODAL INTERVALS.

We mean by dual semantics of modal intervals , their association to the real predicates of PRED(RE) they reject .

DEFINITION 3.1 COPRED(X) := SET( P(.) / -Q(x,X)P(x) )

DEFINITION 3.2 DUAL(A) := INT(SUP(A),INF(A))

Essential theorems in this context are :

THEOREM 3.1 A <* I*(RE) ==> COPRED(A) -= VOID

THEOREM 3.2 COPRED(A) = PRED(RE) - PRED(A)

THEOREM 3.3 A << B <==> DUAL(A) >> DUAL(B)

THEOREM 3.4 P(.) <* COPRED(A) <==> -P(.) <* PRED(DUAL(A))

THEOREM 3.5 A << B <==> COPRED(A) >> COPRED(B)

THEOREM 3.6 IMPR(A) << PROP(A)

THEOREM 3.7 ( A <* Ie(RE) , A -<* Ip(RE) ) <==>
E( P(.) , PRED(RE) ) ( P(.) <* PRED(A) , -P(.) <* PRED(A) )

THEOREM 3.8 For P(.) <* PRED(RE) and A <* I*(RE) ,
one of the two following alternatives holds
(1) ( P(.) <* PRED(PROP(A)) , -P(.) <* PRED(PROP(A)) ) AND
( P(.) <* COPRED(IMPR(A)) , -P(.) <* COPRED(IMPR(A)) )
(2) P(.) <* PRED(IMPR(A)) << PRED(PROP(A)) AND
-P(.) <* COPRED(PROP(A)) << COPRED(IMPR(A))

Perhaps it may be of some use to observe that for A <* I*(RE) , ( A <* Ie(RE) , P(.) <* COPRED(A) ) is equivalent to SET(A) -=* SET(x / P(x)) , and ( A <* Iu(RE) , P(.) <* COPRED(A) ) is equivalent to SET(A) -<< SET(x / P(x)) .

## 4. LATTICE SEMANTICS OF MODAL INTERVALS.

The structure STR( I*(RE) , << ) is isomorphic to the structure STR( CART(RE,RE) ) , CART(>=,<=) ) and , therefore , a distributive lattice like STR( RE , >= ) and STR( RE , <= ) . That is , given A , B <* I*(RE) , their <<-supremum or "join" JOIN( A , B ) , and their <<-infimum or "meet" MEET( A , B ) , do exist , and these operations are mutually distributive with the following operation laws :

```
MEET( A(i) / i <* I )  :=
     ELEM( A /  U(i,I) ( X << A(i) )  <==>  X << A ) =
     INT( MAX( INF(A(i)) / i <* I) , MIN( SUP(A(i)) / i <* I )
JOIN( A(i) / i <* I )  :=
     ELEM( A /  U(i,I) ( X >> A(i) )  <==>  X >> A ) =
     INT( MIN( INF(A(i)) / i <* I) , MAX( SUP(A(i)) / i <* I )
```

Now , for a full identification of the modal intervals A <* I*(RE) with the predicates-set PRED(A), it would be fine that PRED( JOIN(A,B)) would equal UNI(PRED(A),PRED(B)) , and that PRED( MEET(A,B)) wold stand in the same relation towards SEC( PRED(A),PRED(B) ) ; where SEC and UNI stand for the set operations "intersection" and "union" .

This is far from certain , yet not so damaging to prevent a good semantical structure to hold on for the lattice of modal intervals .

To test this property we take , for example , the predicates-set PRED( MEET( INT(1,2),INT(3,4))) = PRED( INT(3,2) ) . The predicate x <* SET( 1.5,3.5 ) belongs to PRED( INT( 1,2) ) and to PRED(INT(3,4)) and therefore to the intersection of these two sets of predicates , but absolutely not to PRED( INT(3,2) ) .

Also , x = 2.5 belongs to PRED( INT(1,4) ) which is equal to PRED( JOIN( INT(1,2) , INT(3,4) ) ) , but neither to PRED( INT(1,2) ) nor to PRED ( INT(3,4) ) .

In terms of this set of predicates, Theorem 2.5 yields the following conclusion :

```
THEOREM 4.1  (1) PRED( MEET(A,B) )  <<  SEC( PRED(A),PRED(B) )
             (2) PRED( JOIN(A,B) )  >>  UNI( PRED(A),PRED(B) )
```

From a structural viewpoint , this theorem , with its "equality failure" , arises from the fact that , if we take

```
DEFINITION 4.1  PRED( I*(RE) ) := SET( PRED(X) / X <* I*(RE) )
```

the system STR( PRED(I*(RE)) , << ) is a sublattice of the larger system STR( PSET(PRED(RE)) , << ) , and the lattice operations MEET and JOIN correspond to the smaller system of the interval-sets of predicates STR( PRED(I*(RE)) , << ) ( PSET is "powerset" ) .

Of course , the result of Theorem 4.1 , failing to provide an equality , stands across the straight on path from the semantics of modal intervals to the semantics of their inclusion-lattice . For a better interpretation of this difficulty , we shall consider , instead of the sets of predicates PRED(X) , some more restricted sets which will provide equality relations replacing the mere inclusions of Theorem 4.1 .

Let us define the sets of :

```
DEFINITION 4.2
     Interval predicates as
        PRED*(RE) := SET( x <* X' / X' <* I(RE) )
     Interval copredicates as
        COPRED*(RE) := SET( x -<* X' / X' <* I(RE) )
     Interval predicates validated ( or accepted ) by A
        PRED*(A) := SET( x <* X' / ( x <* X' ) <* PRED(A) )
```

Interval copredicates covalidated ( or rejected ) by A
COPRED*(A) := SET( x -<* X' / ( x -<* X' ) <* COPRED(A) )
where we say that P(.) is covalidated by A when P(.) <* COPRED(A) .

Now from Theorem 3.4 it follows :

THEOREM 4.2
( x -<* X' ) <* COPRED*(A) <==> ( x <* X') <* PRED*(DUAL(A))

Moreover , the following theorem shows that the belonging relations of ( x <* X' ) and of ( x -<* X' ) , to the sets PRED*(A) and COPRED*(A) , are interval relations indeed :

THEOREM 4.3
(1) ( x <* X' ) <* PRED*(A) <==> IMPR(X') << A
(2) ( x -<* X' ) <* COPRED*(A) <==> PROP(X') >> A
where PROP(X') := ( X' , E )
and IMPR(X') := ( X' , U )

The statement (1) comes out from the left term being equivalent to SET(A) << X' when A is improper, and to SET(A) =* X' when A is proper . Statement (2) results from :
( x -<* X' ) <* COPRED(A) <==>
( x <* X' ) <* PRED(DUAL(A)) <==>
IMPR(X') << DUAL(A) <==>
PROP(X') >> A .

Theorem 4.3 suggests the identifications
( x <* X' ) <-----> IMPR(X')
( x -<* X' ) <-----> PROP(X')
PRED*(A) <-----> SET( IMPR(X') / IMPR(X') << A )
COPRED*(A) <-----> SET( PROP(X') / PROP(X') >> A ) ;
and remark that "point-intervals" X' = INT( x1 ) can be identified to the predicates x = x1 or to the copredicates x -= x1 , according to their conventional membership to the proper or improper class of modal intervals .

Now , from these latter properties , the equalities missing in Theorem 4.1 for PRED( MEET(A,B) ) and PRED( JOIN(A,B) ) , which failed to establish a stronger association between the lattice of intervals and the lattice of interval-sets of predicates , are shown to hold in some cases , but not all , for interval predicates and copredicates :

THEOREM 4.4
(1) PRED*( MEET(A,B) ) = SEC( PRED*(A) , PRED*(B) )
(2) COPRED*( JOIN(A,B) ) = SEC( COPRED*(A) , COPRED*(B) )
(3) PRED*( JOIN(A,B) ) >> UNI( PRED*(A) , PRED*(B) )
(4) COPRED*( MEET(A,B) ) >> UNI( COPRED*(A) , COPRED*(B) )

About (1) , Theorem 4.3.(1) yields inmediatly that ( x <* X' ) <* PRED( MEET(A,B) ) ==> ( x <* X' ) <* SEC( PRED*(A) , PRED*(B) ) . The contrarywise inclusion comes from the lattice property ( IMPR(X') << A , IMPR(X') << B ) ==> IMPR(X') << MEET(A,B) . The assertion (2) is the dual statement of (1) and , moreover , results (3) and (4) are supported by obvious inclusion relations and by Theorems 2.5 and 3.5 . Moreover PRED*( JOIN(A,B) ) can be larger than UNI( PRED*(A) , PRED*(B) ) , as the example of ( x=2.5 ) <* PRED*( JOIN( INT(1,2) , INT(2,4) ) ) = PRED*( INT(1,4) ) shows . All the same , COPRED*( MEET(A,B) ) can be larger than UNI( COPRED*(A),COPRED*(B) ) , as it comes out from the example ( x-=2.5 ) <* COPRED*( MEET(INT(2,1),INT(4,3)) ) = COPRED*( INT(4,1) ).

## 5.- CONCLUDING REMARKS.

Theorems 2.3, 2.5, 2.6, 3.1 and 3.5 , bring out the set-theoretical nature of the inclusion of modal intervals , since they tie modal intervals to the sets of predicates they accept ( validate ) or reject (covalidate) .

Theorems 4.1 and 4.4 , show that the intrinsic structure of the set of modal intervals , with their <<-meet and <<-join operations , does not allow a once for all association of modal intervals , neither with the whole set of the predicates they accept or reject , nor with the more specialized sets of interval-predicates or interval-copredicates .

Modal intervals are , indeed , intrinsically one-sided from the viewpoint of their association with sets of predicates upon the line of real numbers , as they can be identified with the set of interval-predicates they validate , A <----> PRED*(A) , only when interval predicates common to some family of modal intervals SET( A(i) / i<*I ) are to be accounted for , in which case SEC( PRED*( A(i) ) / i<*I ) is equal to PRED*( MEET( A(i) / i<*I ) ) ; and they can be identified with the set of interval copredicates they reject , A <----> COPRED*(A) , only when interval copredicates common to some family of modal intervals do matter , in which case SEC( COPRED*( A(i) ) / i<*I ) is equal to COPRED*( JOIN( A(i) / i<*I ) ) .

Anyway , remark that all the inclusions of Theorems 4.1 and 4.4 become equalities when , between A and B , a relation A << B holds .

An application of the previous theory to the interpretation of interval-rounding results , from the viewpoint of the information they display , is the following theorem :

THEOREM 5.1

If DI << RE is a digital scale for the real numbers , and if outer and inner interval-rounding are defined by

OUT( INT(a,b) ) :=

ELEM( INT(a',b') / a'<*DI , b'<*DI , INT(a',b') >> INT(a,b) )

INN( INT(a,b) ) :=

ELEM( INT(a',b') / a'<*DI , b'<*DI , INT(a',b') << INT(a,b) )

then :

(1) PRED( INN(X) ) << PRED(X)

(2) COPRED( OUT(X) ) << COPRED(X)

(3) If the information supplied by some computing algorithm and/or some observation about a modal interval A is the pair of digital modal intervals A1 , A2 ,with A1 << A << A2 , then , the only predicates and copredicates that are A-decidable "a posteriori" are

the elements of PRED(A1) and of COPRED(A2) .

(4) With the same assumptions as in (3) , the "a priori" information induced by A onto A2 is PRED(A) , and , onto A1 , COPRED(A) .

As a particular application of this theorem to the case of ordinary intervals with the standard outwards rounding A2 >> A , only the "a priori" information PRED(A) ( P(x) with E(x,A)P(x) ==> E(x,A2)P(x) ) and the "a posteriori" information COPRED(A2) ( P(x) with -E(x,A2)P(x) ==> -E(x,A) P(x) , or U(x,A2) -P(x) ==> U(x,A) -P(x) ) are available .

The system of modal intervals can be used for actual computation by using the programming language SIGLA and the simulation language SIMSIGLA developed by the authors .

## 6.- BIBLIOGRAPHY.


Gardeñes E. , Trepat A. : "Fundamentals of SIGLA , an Interval Computing System on the Completed Set of Intervals" Computing 24 , Springer 1980 .

Gardeñes E. , Trepat A. , Janer J.M. : "SIGLA-PL/I , development and applications" Interval Mathematics 1980 , Ed. K. Nickel , Academic Press 1980 .

Gardeñes E. , Trepat A. , Janer J.M. : "Approaches to simulation and to the linear problem in the SIGLA System" Freiburger Intervall-Berichte 81/8 , Freiburg 1982 .

Gardeñes E. , Trepat A. , Mielgo H. : "Present perspective of the SIGLA Interval System" Freiburger Intervall-Berichte 82/9 , Freiburg 1982 .

Gardeñes E. : "Computing with the completed set of intervals : SIGLA-PL/I System" Proc. 22th Science Week , Damascus 1982 ( to appear ) .

Kaucher E. : "Interval analysis in the extended interval space IR" Computing Suppl. 2. ,Springer 1980 .

Nickel K. :"Verbandtheoretische Grundlagen der Intervall-Mathematik" Lecture Notes in Comp. Sc. 29 , Springer 1975 .

# CONVERGENT BOUNDS FOR THE RANGE OF MULTIVARIATE POLYNOMIALS

J. Garloff

Institut für Angewandte Mathematik

Universität Freiburg i.Br.

Freiburg i.Br.

West Germany

## 1. Introduction

In this paper we consider the following problem: We are given a *bivariate polynomial* p, i.e., a polynomial in two variables

$$p(x,y) := \sum_{\mu,\nu=0}^{n} a_{\mu\nu} x^{\mu} y^{\nu} \tag{1.1}$$

having real coefficients $a_{\mu\nu}$ and a rectangle

$$Q = X \times Y \text{ with } X = [\underline{x},\overline{x}],\quad Y = [\underline{y},\overline{y}] \in \mathbb{I}(\mathbb{R}). \tag{1.2}$$

Here $\mathbb{I}(\mathbb{R})$ denotes the set of the compact, nonempty real intervals, henceforth referred to simply as intervals. We are seeking for the *range* of p over Q, i.e.,

$$P(Q) = \{p(x,y) \mid (x,y) \in Q\} = [\underline{m},\overline{m}],$$

$$\text{where } \underline{m} = \min_{(x,y)\in Q} p(x,y),\quad \overline{m} = \max_{(x,y)\in Q} p(x,y).$$

Knowledge of this range, or the equivalent global maximum $\overline{m}$ and global minimum $\underline{m}$ is relevant for numerous investigations and applications in numerical and functional analysis, optimization etc. It is therefore important to find easy and efficient methods for getting good approximations to this range. An exposition of available methods is given in the monograph [10].

In our paper we present two methods for finding convergent upper and lower bounds $\overline{m}_k$, $\underline{m}_k$ for the range $p(Q)$, i.e., $\overline{m}_k \geq \overline{m}$ and $\underline{m}_k \leq \underline{m}$ with $\overline{m}_k \to \overline{m}$ and $\underline{m}_k \to \underline{m}$ for $k \to \infty$. Both methods can easily be extended to the higher-dimensional case. For the sake of simplicity we will present our results only in the bivariate case since the generalization to the higher-dimensional case will be obvious.

The first method is based on the mean value theorem and is given in Section 2. The other method is based on the expansion of a bivariate polynomial in Bernstein polynomials and is discussed in detail in Section 3. Here we also address the problem of finding convergent bounds for the range of p over the unit triangle. It turns out that this can be handled in a similar way as for the rectangle. In Section 4 we consider the case that the coefficients of the polynomial p are not exactly known but can be located between upper and lower bounds

$$a_{\mu\nu} \in A_{\mu\nu} = [\underline{a}_{\mu\nu}, \overline{a}_{\mu\nu}] \in \mathbb{I}(\mathbb{R}), \quad \mu,\nu = 0(1)n. \tag{1.3}$$

Required is now to find the range of a set of bivariate polynomials over Q

$$\left\{ \sum_{\mu,\nu=0}^{n} a_{\mu\nu} x^{\mu} y^{\nu} \;\middle|\; (x,y) \in Q,\ a_{\mu\nu} \in A_{\mu\nu},\ \mu,\nu = 0(1)n \right\}.$$

We conclude our paper with a particular application to a problem in multidimensional system theory, namely testing a bivariate polynomial for positivity.

Each real interval can be mapped onto $[0,1]$ by a linear function. So we will confine our discussion mainly to the *unit square* $I := [0,1]\times[0,1]$. For an integer k we define $K := \{(i,j) \mid i,j = 0(1)k\}$.

## 2. Bounds using function values

The following is an extension of a method developed by Rivlin [12] for the univariate case. The bounds involve the function values of the polynomial on the grid on the unit square I given by $(\mu/k, \nu/k)$, $(\mu,\nu) \in K$.

Theorem 1 (without proof):
Let p be given by (1.1). Then

$$\begin{aligned}\max_{(x,y)\in I} p(x,y) &= \overline{m} \leqq \max_{(\mu,\nu)\in K} p\left(\frac{\mu}{k},\frac{\nu}{k}\right) + \alpha_k \\ \min_{(x,y)\in I} p(x,y) &= \underline{m} \geqq \min_{(\mu,\nu)\in K} p\left(\frac{\mu}{k},\frac{\nu}{k}\right) - \alpha_k ,\end{aligned} \tag{2.1}$$

where

$$\alpha_k := \frac{1}{8k^2} \sum_{i,j=0}^{n} (i+j)(i+j-1)|a_{ij}| .$$

Remark: If p has a large number of vanishing coefficients it might be advantageous to apply the algorithm given in [11] to evaluate p at the grid points since this algorithm takes account of the sparsity pattern of p.

## 3. Bounds using the Bernstein form

In this section we derive bounds for the range of a bivariate polynomial (1.1) on the unit square using the so called Bernstein form. This form is intimately related to Bernstein polynomials (a good reference for Bernstein polynomials is the monograph [9]). The first application to the range of univariate polynomials was given by Cargo and Shisha [5]; Rivlin [12] improved upon the bounds obtained by Cargo and Shisha. Grassmann and Rokne [6] and Rokne [13-16] applied the results of Rivlin to real and complex interval polynomials. Finally, Lane and Riesenfeld [8] discussed subdivision in the univariate case.

### 3.1 The Bernstein form of a bivariate polynomial on the unit square

Let $k \geqq n$ be an integer. We define for $(i,j) \in K$.

$$p_{ij}^{(k)}(x,y) := \binom{k}{i}\binom{k}{j}x^i(1-x)^{k-i}y^j(1-y)^{k-j},\quad x,y \in I. \tag{3.1}$$

Then by some manipulations we get the identity

$$x^\mu y^\nu = \sum_{s=\mu,t=\nu}^{k} \binom{s}{\mu}\binom{t}{\nu}\rho_{\mu\nu}^{(k)} p_{st}^{(k)}(x,y), \tag{3.2}$$

$$\text{where } \rho_{\mu\nu}^{(k)} := \left[\binom{k}{\mu}\binom{k}{\nu}\right]^{-1},\quad (\mu,\nu) \in K. \tag{3.3}$$

Substituting (3.2) into (1.1) gives

$$p(x,y) = \sum_{(i,j)\in K} b_{ij}^{(k)} p_{ij}^{(k)}(x,y), \tag{3.4}$$

$$\text{where } b_{ij}^{(k)} := \sum_{s=0}^{i}\sum_{t=0}^{j} \binom{i}{s}\binom{j}{t}\rho_{st}^{(k)} a_{st}, \tag{3.5}$$

with the convention that $a_{st} = 0$ for $s > n$ or $t > n$.

We call the $b_{ij}^{(k)}$ the *Bernstein coefficients* and (3.4) the *Bernstein form* of p (on the unit square).

Theorem 2:
If p is given by (1.1), then we have

$$\max_{(i,j)\in K} b_{ij}^{(k)} \geqq \bar{m},\quad \underline{m} \geqq \min_{(i,j)\in K} b_{ij}^{(k)} \tag{3.6}$$

for each $k \geqq n$; equality holds in the left (resp., right) inequality if and only if $\max_{(i,j)\in K} b_{ij}^{(k)}$ (resp., $\min_{(i,j)\in K} b_{ij}^{(k)}$) is one of $b_{00}^{(k)}$, $b_{k0}^{(k)}$, $b_{0k}^{(k)}$, $b_{kk}^{(k)}$.

Proof: Since $0 \leqq p_{ij}^{(k)}(x,y)$ for all $(x,y) \in I$ and $(i,j) \in K$, and

$$\sum_{(i,j)\in K} p_{ij}^{(k)}(x,y) = 1 \quad \text{for all } (x,y) \in I \tag{3.7}$$

the inequalities (3.6) follow.
The "if" part is obvious from

$$b_{00}^{(k)} = a_{00} = p(0,0),\quad b_{0k}^{(k)} = \sum_{t=0}^{k} a_{0t} = p(0,1),$$

$$b_{k0}^{(k)} = \sum_{s=0}^{k} a_{s0} = p(1,0), \quad b_{kk}^{(k)} = \sum_{(s,t)\in K} a_{st} = p(1,1).$$

We now assume that $\max b_{ij}^{(k)} = \overline{m} = p(\hat{x},\hat{y})$, $(\hat{x},\hat{y}) \in I$, and $\max b_{ij}^{(k)} > b_{00}^{(k)}, b_{0k}^{(k)}, b_{k0}^{(k)}, b_{kk}^{(k)}$. If $\hat{x},\hat{y} \in (0,1)$, then by (3.7)

$$p(\hat{x},\hat{y}) < \max b_{ij}^{(k)} \sum_{(i,j)\in K} p_{ij}(\hat{x},\hat{y}) = \max b_{ij}^{(k)},$$

a contradiction. The proof of the other cases and for $\min b_{ij}^{(k)}$ is analogous.

∎

We now show that the bounds given in (3.6) converge to $\underline{m}$ and $\overline{m}$, respectively.

<u>Theorem 3</u>:
If $k \geqq 2$, then

$$\max_{(i,j)\in K} b_{ij}^{(k)} - \overline{m}, \ \underline{m} - \min_{(i,j)\in K} b_{ij}^{(k)} \leqq \gamma(k-1)k^{-2},$$

where

$$\gamma := \sum_{\mu,\nu=0}^{n} ((\mu-1)_+^2 + (\nu-1)_+^2)\,|a_{\mu\nu}| \tag{3.8}$$

and $(x)_+ = \max(0,x)$.

<u>Proof</u>: Since some of our considerations follow Rivlin's proof for the univariate case [12] we only give an outline of the proof.

For a function f defined on I let

$$B_k(f;x,y) := \sum_{(i,j)\in K} f(\tfrac{i}{k},\tfrac{j}{k})\,p_{ij}^{(k)}(x,y).$$

For $s,t \leqq n$, denote by $\delta_{ij}(s,t)$, $(i,j) \in K$, the Bernstein coefficients of the polynomial $B_k(x^s x^t;x,y) - x^s y^t$. Since $B_k(x^s y^t;x,y) = x^s y^t$ for $s,t \leqq 1$ we have $\delta_{ij}(s,t) = 0$ for $(i,j) \in K$ and $s,t \leqq 1$. Therefore, we assume that $s \geqq 2$ or $t \geqq 2$.
If $0 \leqq i < s$ we have by (3.2)

$$\delta_{ij}(s,t) \leqq (\tfrac{i}{k})^s(\tfrac{j}{k})^t \leqq (\tfrac{i}{k})^s \leqq (\tfrac{s-1}{k})^s \leqq \frac{(s-1)^2}{k}(1-\tfrac{1}{k}).$$

If $2 \leq s \leq i$ and $2 \leq t \leq j$ we get after some algebraic manipulations

$$\delta_{ij}(s,t) = (\tfrac{i}{k})^s (\tfrac{j}{k})^t - \rho_{st}^{(k)} \binom{i}{s}\binom{j}{t}$$

$$\leq (\tfrac{i}{k})^s (\tfrac{j}{k})^t \left[1 - \left(1 - \frac{(s-1)}{i}\right)^{s-1}\left(1 - \frac{(t-1)}{j}\right)^{t-1}\right] .$$

Now we apply the generalized Bernoulli inequality, see e.g. [7, p. 60], to obtain

$$\delta_{ij}(s,t) \leq (\tfrac{i}{k})^s (\tfrac{j}{k})^t \left( \frac{(s-1)^2}{i} + \frac{(t-1)^2}{j} \right)$$

$$= (\tfrac{i}{k})^{s-1} (\tfrac{j}{k})^t \frac{(s-1)^2}{k} + (\tfrac{i}{k})^s (\tfrac{j}{k})^{t-1} \frac{(t-1)^2}{k} .$$

It follows that (note that $\delta_{kk}(s,t) = 0$)

$$\delta_{ij}(s,t) \leq \frac{k-1}{k^2} ((s-1)^2 + (t-1)^2).$$

It is easy to see that this formula is also true in the remaining cases.
As in the univariate case now one shows that

$$\left| p(\tfrac{i}{k}, \tfrac{j}{k}) - b_{ij}^{(k)} \right| \leq \gamma \frac{k-1}{k^2}$$

from which the assertion follows. ⊠

Because of Theorem 3 one expects that when increasing k the bounds become better. Before we discuss this in more detail we note another improvement, namely a correction of already calculated bounds.
We assume that

$$\max_{(i,j)\in K} b_{ij}^{(k)} = b_{\mu\nu}, \quad (\mu,\nu) \notin \{(0,0), (0,k), (k,0), (k,k)\},$$

$$\max \{b_{ij}^{(k)} \mid (i,j) \neq (\mu,\nu)\} = b_{\hat{\mu},\hat{\nu}} < b_{\mu\nu}.$$

By a similar argument as in [12] one shows using

$$\max_{(x,y)\in I} p_{ij}(x,y) = p_{ij}(\tfrac{i}{k}, \tfrac{j}{k})$$

that

$$\overline{m} \leq p_{\mu\nu}(\tfrac{\mu}{k},\tfrac{\nu}{k})b_{\mu\nu} + (1 - p_{\mu\nu}(\tfrac{\mu}{k},\tfrac{\nu}{k}))b_{\hat{\mu}\hat{\nu}} < b_{\mu\nu} . \qquad (3.9)$$

If the maximum of the Bernstein coefficients is assumed more than once then a similar bound holds involving the maxima of the corresponding Bernstein polynomials.
An analogous bound is valid for $\underline{m}$.

## 3.2 Calculation of the Bernstein coefficients

The calculation of the Bernstein coefficients by (3.5) is not economic since, e.g., the number of additions needed is $\frac{1}{4}n^4 + O(n^3)$ $(k = n)$. We present now a method for calculating the Bernstein coefficients which requires fewer arithmetical operations.

Proposition 1:
For $\mu,\nu = 0(1)n$ we have

$$a_{\mu\nu} = (\rho_{\mu\nu}^{(k)})^{-1}\Delta_{\mu\nu}b_{00}^{(k)} ,$$

where $\Delta_{\mu\nu}$ is a twodimensional forward difference operator defined by

$$\Delta_{\mu\nu}b_{ij}^{(k)} := \sum_{\sigma=0}^{\mu}\sum_{\tau=0}^{\nu}(-1)^{\mu+\nu-\sigma-\tau}\binom{\mu}{\sigma}\binom{\nu}{\tau}b_{i+\sigma,j+\tau}^{(k)} ,$$

$$\mu \leq k-i, \ \nu \leq k-j .$$

Proof: Straightforward calculation using

$$a_{st} = \frac{1}{s!t!}\frac{\partial^{s+t}p}{\partial x^s \partial y^t}(0,0) . \qquad \boxtimes$$

To calculate the Bernstein coefficients one may proceed in two steps:

First, one computes

$$\Delta_{\mu\nu}b_{00}^{(k)} = \rho_{\mu\nu}^{(k)}a_{\mu\nu} .$$

Then one computes the Bernstein coefficients from $\Delta_{\mu\nu}b_{00}^{(k)}$ by using the following recurrence relations:

$$\left.\begin{aligned}
\Delta_{00} b_{ij}^{(k)} &= b_{ij}^{(k)} , \\
\Delta_{10} b_{ij}^{(k)} &= b_{i+1,j}^{(k)} - b_{ij}^{(k)} , \\
\Delta_{01} b_{ij}^{(k)} &= b_{i,j+1}^{(k)} - b_{ij}^{(k)} , \\
\Delta_{\mu+1,\nu} b_{ij}^{(k)} &= \Delta_{\mu\nu} b_{i+1,j}^{(k)} - \Delta_{\mu\nu} b_{ij}^{(k)} , \\
\Delta_{\mu,\nu+1} b_{ij}^{(k)} &= \Delta_{\mu\nu} b_{i,j+1}^{(k)} - \Delta_{\mu\nu} b_{ij}^{(k)} ;
\end{aligned}\right\} \qquad (3.10)$$

furthermore, we have

$$\Delta_{\mu\nu} b_{00}^{(k)} = 0 \quad \text{if} \quad \mu > n \quad \text{or} \quad \nu > n .$$

To clarify the second step we give the explicit calculations in the case $k = n = 2$ (for simplicity we suppress the upper index k). We start with the table of the differences $\Delta_{\mu\nu} b_{00}$

$$\begin{array}{c|ccc}
b_{00} & \Delta_{01} b_{00} & \Delta_{02} b_{00} \\ \hline
\Delta_{10} b_{00} & \Delta_{11} b_{00} & \Delta_{12} b_{00} \\
\Delta_{20} b_{00} & \Delta_{21} b_{00} & \Delta_{22} b_{00}
\end{array} \; ;$$

in the upper left corner we have $b_{00} = \Delta_{00} b_{00}$ .

Now add the first column to the second and the second to the third to obtain

$$\begin{array}{cc|c}
b_{00} & b_{01} & \Delta_{01} b_{01} \\ \hline
\Delta_{10} b_{00} & \Delta_{10} b_{01} & \Delta_{11} b_{01} \\
\Delta_{20} b_{00} & \Delta_{20} b_{01} & \Delta_{21} b_{01}
\end{array} \; ;$$

as the second Bernstein coefficient we now know $b_{01} = \Delta_{00} b_{01}$. Now add row 1 to row 2 and row 2 to row 3 which gives

$$\begin{array}{cc|c}
b_{00} & b_{01} & \Delta_{01} b_{01} \\
b_{10} & b_{11} & \Delta_{01} b_{11} \\ \hline
\Delta_{10} b_{10} & \Delta_{10} b_{11} & \Delta_{11} b_{11}
\end{array} \; .$$

In the last but one step add the second column to the third to obtain

$$\begin{array}{ccc} b_{00} & b_{01} & b_{02} \\ b_{10} & b_{11} & b_{12} \\ \hline \Delta_{10}b_{10} & \Delta_{10}b_{11} & \Delta_{10}b_{12} \end{array}$$

and in the last step the second row to the third which yields the table of the Bernstein coefficients $b_{ij}$, $i,j = 0(1)2$.

The number of additions required for this method is $n(n+1)^2$ $(k = n)$. Also the number of calculations of binomial coefficients is considerably smaller than for the direct calculation by (3.5).

When the Bernstein coefficients are computed by the difference table for several k then for each k all the Bernstein coefficients have to be calculated once again. Hence the difference table is unfavourable when it is used more than once. A better way is to calculate the Bernstein coefficients for fixed k-1 and then to make use of the following recurrence relations ($(i,j) \in K$):

$$\begin{aligned} b_{ij}^{(k)} &= k^{-2}[ij\, b_{i-1,j-1}^{(k-1)} + j(k-i)b_{i,j-1}^{(k-1)} \\ &\quad + i(k-j)b_{i-1,j}^{(k-1)} + (k-i)(k-j)b_{ij}^{(k-1)}] \qquad (3.11) \\ &\quad \text{with } b_{-1,-1}^{(k-1)} = b_{-1,j}^{(k-1)} = b_{i,-1}^{(k-1)} = b_{ik}^{(k-1)} = b_{kj}^{(k-1)} = 0, \\ &\quad i,j = 0(1)k. \end{aligned}$$

We see that the Bernstein coefficients of order k are convex linear combinations of Bernstein coefficients of order k-1 and we may conclude that the convergence of the bounds is monotone:

$$\max b_{ij}^{(k-1)} \geqq \max b_{ij}^{(k)} ,$$

$$\min b_{ij}^{(k-1)} \leqq \min b_{ij}^{(k)} .$$

## 3.3 Subdivision

In this section we discuss subdivision, i.e., dividing the unit square I into four subsquares of edge length 1/2 (see Fig. 1)

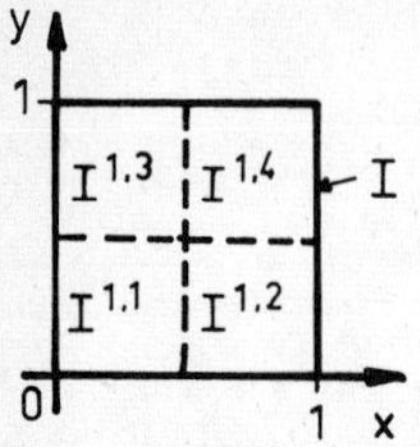


Fig. 1. Subdivision

and calculating the Bernstein coefficients of the given polynomial (1.1) on each subsquare. Here we mean by the Bernstein coefficients $b_{ij}(Q)$ of p on a rectangle Q given by (1.2) the Bernstein coefficients of the shifted polynomial

$$p(x,y) = \sum_{\mu,\nu=0}^{n} c_{\mu\nu}\xi^{\mu}\eta^{\nu} \ , \ (\xi,\eta) \in I \tag{3.12}$$

with

$$c_{\mu\nu} = (\bar{x}-\underline{x})^{\mu}(\bar{y}-\underline{y})^{\nu} \sum_{s=\mu}^{n} \sum_{t=\nu}^{n} \binom{s}{\mu}\binom{t}{\nu}\underline{x}^{s-\mu}\underline{y}^{t-\nu}a_{st} \ . \tag{3.13}$$

The coefficients $c_{\mu\nu}$ of the shifted polynomial may be calculated by a twodimensional complete Horner scheme. The process may be continued by subdividing again each of the four subsquares into four subsquares of edge length 1/4 and calculating the Bernstein coefficients on each of the 16 subsquares and so on. Then the maximum (resp., minimum) taken over the Bernstein coefficients of p on all subsquares is an upper (resp., lower) bound for p over I.

We first give the explicit formulas for the Bernstein coefficients on the four subsquares of an arbitrary square. By iterated use of these formulas one sees that the Bernstein coefficients on the subsquares generated by subdivision may be calculated successively from the Bernstein coefficients on I. In particular, transformation of the subsquares onto I can be avoided. Then we show that the bounds converge quadratically when subdivision is applied iteratively.

In the sequel we assume that the Bernstein coefficients of p are computed for fixed k, and we suppress the upper index k.

Proposition 2:

Let the rectangle $Q$ be given by (1.2) and let $\xi := (\underline{x}+\overline{x})/2$, $\eta := (\underline{y}+\overline{y})/2$. Then the Bernstein coefficients on the four subrectangles are given by $((i,j) \in K)$:

$$b_{ij}(Q^{\ell}) = 2^{-i-j} \sum_{s=0}^{i} \sum_{t=0}^{j} \binom{i}{s}\binom{j}{t} \beta_{st}^{(\ell)} , \qquad \ell = 1,2,3,4,$$

where

$$\beta_{st}^{(1)} = b_{st}(Q) \qquad \text{on } Q^1 := [\underline{x},\xi]\times[\underline{y},\eta] ,$$

$$\beta_{st}^{(2)} = b_{k-s,t}(Q) \qquad \text{on } Q^2 := [\xi,\overline{x}]\times[\underline{y},\eta] ,$$

$$\beta_{st}^{(3)} = b_{s,k-t}(Q) \qquad \text{on } Q^3 := [\underline{x},\xi]\times[\eta,\overline{y}] ,$$

$$\beta_{st}^{(4)} = b_{k-s,k-t}(Q) \qquad \text{on } Q^4 := [\xi,\overline{x}]\times[\eta,\overline{y}] .$$

Proof: Similar as for subdivision in the univariate case [8]. ⊠

For practical calculation of the Bernstein coefficients $b_{ij}(Q^{\ell})$, $\ell = 1,2,3,4$, one writes down the following four tables

$$(b_{ij}), \quad (b_{k-i,j}), \quad (b_{i,k-j}), \quad (b_{k-i,k-j})$$

and then proceeds for each table similarly as in the calculation of the Bernstein coefficients, cf. Section 3.2. The entries of the final tables have to be divided by $2^{-i-j}$. We see that the $b_{ij}(Q^{\ell})$ are convex linear combinations of the $b_{ij}(Q)$. Therefore, we conclude that the bounds calculated from the Bernstein coefficients on the smaller rectangles are at least as good as those obtained by using the Bernstein coefficients of p on Q and that the bounds are monotone when subdivision is applied iteratively.

We denote the subsquares of edge length $2^{-m}$ generated by subdivision of I (arranged in any specified order) by $I^{m,\ell}$ , $\ell = 1(1)4^m$.

Theorem 4:
If $k \geq 2$ the following relation hold for all $m = 3,4,5,\dots$

$$\left.\begin{array}{l} \max\limits_{\substack{(i,j)\in K\\ \ell=1(1)4^m}} b_{ij}(I^{m,\ell}) - \bar{m} \\[2ex] \bar{m} - \min\limits_{\substack{(i,j)\in K\\ \ell=1(1)4^m}} b_{ij}(I^{m,\ell}) \end{array}\right\} \leq \varepsilon(k-1)k^{-2}m^{-2},$$

where

$$\varepsilon := \max_{\mu,\nu=0(1)n} \sum_{s=\mu}^{n} \sum_{t=\nu}^{n} \binom{s}{\mu}\binom{t}{\nu} |a_{st}| \; .$$

Proof: Let m be fixed and

$$\max_{\substack{(i,j)\in K\\ \ell=1(1)4^m}} b_{ij}(I^{m,\ell}) = \max_{(i,j)\in K} b_{ij}(I^{m,\ell_o}) \; .$$

Then by Theorem 3

$$\max_{(i,j)\in K} b_{ij}(I^{m,\ell_o}) - \max_{(x,y)\in I^{m,\ell_o}} p(x,y) \leq \gamma(k-1)k^{-2} \; ,$$

where

$$\gamma = \sum_{\mu,\nu=1}^{n-1} (\mu^2+\nu^2)|c_{\mu+1,\nu+1}|$$

$$+ \sum_{\mu=2}^{n} (\mu-1)^2 (|c_{\mu o}|+|c_{\mu 1}|+|c_{o\mu}|+|c_{1\mu}|) \; ,$$

and the $c_{\mu\nu}$'s are the coefficients of the shifted polynomial p (3.12). Since by (3.13)

$$|c_{\mu\nu}| \leq \varepsilon\cdot 2^{-m(\mu+\nu)} \quad , \; (\mu,\nu) \in K$$

we get (we assume w.l.o.g. $p \neq 0$)

$$\gamma\varepsilon^{-1} \leqq 2^{-2m} \sum_{\mu,\nu=1}^{n-1} (\mu^2+\nu^2) 2^{-m(\mu+\nu)} + 2^{-m+1}(1+2^{-m}) \sum_{\mu=1}^{n-1} \mu^2 2^{-m\mu} .$$

Because all summands on the right hand side are positive we may estimate the sums by the respective infinite series. Making use of the relation

$$\sum_{\mu=1}^{\infty} \mu^2 x^{\mu} = x(1+x)(1-x)^{-3} \quad \text{for} \quad |x| < 1$$

we obtain

$$\gamma\varepsilon^{-1} \leqq 2^{-4m+1}(1+2^{-m})(1-2^{-m})^{-4} + 2^{-2m+1}(1+2^{-m})^2(1-2^{-m})^{-3}$$

$$= 2^{-2m+1}(1+2^{-m})(1-2^{-m})^{-4}$$

$$= 2^{m+1}(2^m+1)(2^m-1)^{-4} .$$

The last term is less than $m^{-2}$ for $m \geqq 3$.
The proof for the minimum is entirely analogous. ⊠

Remarks: i) One shows similarly that subdivision in the univariate case converges quadratically. This extends the results in [8].

ii) For the proof of quadratic convergence in the r-variate case with $r \geqq 3$ it is easier to use the rougher estimate

$$\gamma \leqq \varepsilon \sum_{\mu_1,\ldots,\mu_r=0}^{\infty} (\mu_1^2 + \ldots + \mu_r^2) 2^{-m(\mu_1+\ldots+\mu_r)} .$$

For the question of which subsquares can be discarded from the list of further examination see Section 5.

## 3.4 Bounds for the range of a multivariate polynomial on the unit triangle

In this section we consider the *unit triangle*

$$S := \{(x,y) \in \mathbb{R}^2 \mid x,y \geqq 0 \wedge x+y \leqq 1\}$$

instead of the unit square and address the question of finding bounds for the range $[\underline{m},\overline{m}]$ of the bivariate polynomial p given by (1.1) on S, i.e.,

$$\underline{m} = \min_{(x,y)\in S} p(x,y), \quad \overline{m} = \max_{(x,y)\in S} p(x,y).$$

We set

$$d := \max\{\mu+\nu \mid a_{\mu\nu} \neq 0\}, \tag{3.14}$$

e.g., $d = 2n$ if $a_{nn} \neq 0$, $d = n$ if $a_{\mu\nu} = 0$ for $\nu > n-\mu$. Let $k \geqq d$. The index set K is now defined by

$$K := \{(i,j) \mid i,j = 0(1)k,\ i+j \leqq k\}.$$

Appropriate bivariate Bernstein polynomials are now given by $((i,j)\in K)$:

$$p_{ij}^{(k)}(x,y) = \binom{k}{i\ j} x^i y^j (1-x-y)^{k-i-j}$$

with the generalized binomial coefficients

$$\binom{k}{i\ j} := \frac{k!}{i!j!(k-i-j)!} = \binom{k}{i}\binom{k-i}{j}.$$

Then on the unit triangle S we have again the Bernstein form (3.4) with the Bernstein coefficients (3.5) but now with the convention that

$$\rho_{\mu\nu}^{(k)} := \binom{k}{\mu\ \nu}^{-1} \quad \text{for all} \quad (\mu,\nu) \in K \tag{3.3'}$$

instead of (3.3).
For these Bernstein polynomials Theorem 2 remains true with the modification that in the case of equality $\max b_{ij}^{(k)}$, $\min b_{ij}^{(k)}$ $\in \{b_{00}^{(k)}, b_{0k}^{(k)}, b_{k0}^{(k)}\}$. Theorem 3 reads now (due to the fact that $\delta_{\mu\nu}(1,1) = -\mu\nu\, k^{-2}(k-1)^{-1}$, cf. the proof of Theorem 3):

<u>Theorem 3'</u>:
The following bound holds for all $k \geqq 2$

$$\max_{(i,j)\in K} b_{ij}^{(k)} - \overline{m},\ \underline{m} - \min_{(i,j)\in K} b_{ij}^{(k)} \leqq (\gamma + |a_{11}|)(k-1)k^{-2},$$

where $\gamma$ is given by (3.8).

Also, already calculated bounds may be improved similarly as for the case of the unit square, cf. (3.9). The calculation of the Bernstein coefficients on I carries over to the Bernstein coefficients on S but now only the upper left half of the difference table is needed. The number of additions required is $(k=n)$ $d(d+1)(d+2)/3$, where d is given by (3.14). Formula (3.11) reads now $((i,j) \in K)$:

$$b_{ij}^{(k)} = k^{-1}[i\, b_{i-1,j}^{(k-1)} + j\, b_{i,j-1}^{(k-1)} + (k-i-j) b_{ij}^{(k-1)}] \qquad (3.11')$$

$$\text{with } b_{-1,j}^{(k-1)} = b_{i,-1}^{(k-1)} = 0,\ b_{ij}^{(k-1)} = 0 \text{ if } i+j = k.$$

Results on subdivision will be given elsewhere.

### 3.5 Symmetric coefficients

It is advantageous if the coefficients of p given by (1.1) are *symmetric*, i.e., $a_{\mu\nu} = a_{\nu\mu}$, $\mu,\nu = 0(1)n$, because then the Bernstein coefficients (on the unit square and the unit triangle) are also symmetric and therefore the number of operations required for the calculation of the Bernstein coefficients can about be halved. The symmetry of the Bernstein coefficients on I carries over to the Bernstein coefficients on the subsquares generated by subdivision but in different forms. So we have for all $(i,j) \in K$ (cf. Figure 1)

$$b_{ij}^{(k)}(I^{1,\ell}) = b_{ji}^{(k)}(I^{1,\ell}) \ , \ \ell = 1,4$$

$$b_{ij}^{(k)}(I^{1,2}) = b_{ji}^{(k)}(I^{1,3})$$

and again the number of operations required can about be halved.

## 4. Bounds for the range of a multivariate interval polynomial

To guarantee the bounds obtained in Sections 2 and 3 in the presence of rounding errors (entailed by computing with fixed length floating point arithmetic) interval arithmetic, see, e.g., [1], should be applied. In this section we address another question involving interval arithmetic.

Assume that the coefficients $a_{\mu\nu}$ of the polynomial p given by (1.1) are not exactly known but can be located between upper and lower bounds (1.3). Then we have to consider the *interval polynomial*

$$P(x,y) := \sum_{\mu,\nu=0}^{n} A_{\mu\nu} x^{\mu} y^{\nu}$$

and it is required to find the range of P over the unit square I, $P(I) = \{P(x,y) \mid (x,y) \in I\}$. Clearly, $p(I) \subseteq P(I)$ holds. Bounds for P(I) can be obtained from (2.1), (3.6) simply by replacing the real coefficients $a_{\mu\nu}$ by the respective interval coefficients $A_{\mu\nu}$ and the real arithmetical operations by the respective interval arithmetical operations. E.g., an enclosure of the Bernstein coefficients $b_{ij}^{(k)}$ is given by

$$B_{ij}^{(k)} = [\underline{b}_{ij}^{(k)}, \overline{b}_{ij}^{(k)}] := \sum_{s=0}^{i} \sum_{t=0}^{j} \binom{i}{s}\binom{j}{t} \rho_{st}^{(k)} A_{st}, \quad (i,j) \in K; \quad (4.1)$$

then max $\overline{b}_{ij}^{(k)}$ (resp., min $\underline{b}_{ij}^{(k)}$) is an upper (resp., lower) bound for max P(I) (resp., min P(I)). We do not go into the details here. However, special attention has to be paid to the overestimation entailed by replacing real numbers by intervals. This concerns two problems which we will discuss in the sequel.

1) In (4.1) each coefficient $A_{\mu\nu}$ occurs only once in the calculation of each $B_{ij}^{(k)}$. It follows that the direct calculation by (4.1) is *optimal*, i.e.,there is a real polynomial $\tilde{p}(x,y) = \sum_{\mu,\nu=0}^{n} \tilde{a}_{\mu\nu} x^{\mu} y^{\nu}$ with $\tilde{a}_{\mu\nu} \in A_{\mu\nu}$ and Bernstein coefficients $\tilde{b}_{ij}^{(k)}$ such that

$$\max \tilde{b}_{ij}^{(k)} = \max \overline{b}_{ij}^{(k)} \geq \max_{(x,y)\in I} P(x,y)$$

and analogously for min $\underline{b}_{ij}^{(k)}$.

An enclosure of the real Bernstein coefficients $b_{ij}^{(k)}$ may also be obtained by interval performance of the procedure using the difference table, cf. Section 3.2, starting with $\rho_{\mu\nu}^{(k)} A_{\mu\nu}$. In the calculation of each entry of the resulting table some $A_{\mu\nu}$'s occur more than once. But this causes no overestimation because the real numbers by which the intervals under consideration are multiplied are positive and therefore the distributive law holds, see, e.g., [1, p. 3]. It follows that also the difference table produces the coefficients $B_{ij}^{(k)}$.

2) If the range of P over an arbitrary rectangle Q given by (1.2) is wanted the interval polynomial has to be shifted to I. Then when replacing in (3.13) all real coefficients $a_{\mu\nu}$ by the intervals $A_{\mu\nu}$ the width of the $B_{ij}^{(k)}(Q)$ could increase since each of the intervals $A_{\mu\nu}$ occurs more than once in the calculation of each $B_{ij}^{(k)}(Q)$. But if one plugs (3.13) into (3.5) then the resulting double sum can be rearranged as follows (for the univariate case see [16]):

$$b_{ij}^{(k)} = \sum_{s,t=0}^{n} a_{st} \sum_{\mu=0}^{\min\{s,n,i\}} \binom{s}{\mu} \underline{x}^{s-\mu} (\overline{x}-\underline{x})^{\mu} \frac{\binom{i}{\mu}}{\binom{k}{\mu}} \times$$

$$\times \sum_{\nu=0}^{\min\{t,n,j\}} \binom{t}{\nu} \underline{y}^{t-\nu} (\overline{y}-\underline{y})^{\nu} \frac{\binom{j}{\nu}}{\binom{k}{\nu}} . \qquad (4.2)$$

Now each real number $a_{\mu\nu}$ occurs only once in the calculation of each $b_{ij}^{(k)}$ and replacing $a_{st}$ by $A_{st}$ in (4.2) gives an optimal formula, i.e., the endpoints of the resulting intervals are Bernstein coefficients of real polynomials with coefficients taken from the interval coefficients. However, this formula requires more calculations compared to (4.1).

Another way to avoid the overestimation entailed by the shift of the original polynomial is to divide the given rectangle into (at most four) rectangles lying in the four quadrants of $\mathbb{R}^2$. On each of these rectangles the two *corner polynomials* of P, i.e.,

$$\sum_{\mu,\nu=0}^{n} a_{\mu\nu}^{(1)} x^{\mu} y^{\nu} \leqq \sum_{\mu,\nu=0}^{n} \tilde{a}_{\mu\nu} x^{\mu} y^{\nu} \leqq \sum_{\mu,\nu=0}^{n} a_{\mu\nu}^{(2)} x^{\mu} y^{\nu}$$

$$\text{for all } \tilde{a}_{\mu\nu} \in A_{\mu\nu} \quad \text{with } a_{\mu\nu}^{(1)}, a_{\mu\nu}^{(2)} \in A_{\mu\nu}$$

can be given explicitly, e.g. on $[0,\infty)\times(-\infty,0]$ we have

$$a_{\mu\nu}^{(1)} = \underline{a}_{\mu\nu} \text{ if } \nu \text{ is even}, = \overline{a}_{\mu\nu} \text{ if } \nu \text{ is odd}$$

$$a_{\mu\nu}^{(2)} = \overline{a}_{\mu\nu} \text{ if } \nu \text{ is even}, = \underline{a}_{\mu\nu} \text{ if } \nu \text{ is odd}.$$

Then the problem of finding bounds for the range of the interval polynomial P reduces to the problem of finding bounds for the range of at most 8 real polynomials for which the shift to I can be done exactly (except for small intervals in the coefficients of the shifted polynomial due to the use of machine interval arithmetic).

## 5. An application

This paper was stimulated by a problem often arising in multidimensional system theory, namely to test a multivariate polynomial p for positivity on a multidimensional rectangle, cf. [2 ; 3, Chapter 2, 17]. Such tests are often referred to as *local positivity tests*.
Computing the minimum of the Bernstein coefficients provides an alternative local positivity test. If $\min_{(i,j)\in K} b_{ij}^{(k)} > 0$ for an integer k then the positivity of p is guaranteed. On the other hand, if

$$\min_{(i,j)\in K} b_{ij}^{(k)} + \gamma(k-1)k^{-2} \leqq 0$$

then by Theorem 3, p assumes also nonpositive values.
If subdivision is applied iteratively one may reduce the computational effort by the following two observations. Let $I^* = I^{m,\ell}$ be a subsquare generated by subdivision.

i) If the polynomial p assumes its minimum over $I^*$ at one of the four vertices of $I^*$, then by Theorem 2

$$\min_{(x,y)\in I^*} p(x,y) = \min_{(i,j)\in K} b_{ij}(I^*) \in \{b_{00}(I^*), b_{k0}(I^*), b_{0k}(I^*), b_{kk}(I^*)\}$$

and there is no need to subdivide $I^*$ further. If furthermore

$$\min_{(i,j)\in K} b_{ij}(I^*) = \min_{\substack{(i,j)\in K \\ \ell=1(1)4^m}} b_{ij}(I^{m,\ell})$$

then one already knows that

$$\min_{(x,y)\in I^*} p(x,y) = \min_{(x,y)\in I} p(x,y) .$$

ii) If

$$\min_{(i,j)\in K} b_{ij}(I^*) > \min_{\substack{(i,j)\in K \\ \ell=1(1)4^m}} b_{ij}(I^{m,\ell}) + \varepsilon(k-1)k^{-2}m^{-2} \qquad (5.1)$$

then $I^*$ may be discarded from the list for further examination since p can not assume its minimum over I on $I^*$. If equality holds in (5.1) $I^*$

may also be discarded since the range of p over I* makes no additional contribution to the minimum of p over I.

Multivariate polynomials can be very sensitive to small perturbations of their coefficients. It is therefore useful to know the allowable intervals centered around the respective unperturbed values within the coefficients of the polynomial might fluctuate without losing the property of being positive for all $x \in \mathbb{R}^r$. This problem was solved in [4 ; see also 3, Section 2.5].

## References


[1] Alefeld, G. and J. Herzberger, Introduction to Interval Computations, Academic Press, New York (1983)

[2] Bickard, T.A. and E.I. Jury, Real polynomials: a test for non-global non-negativity and non-global positivity, J. Math. Anal. Appl. 78, 17-32 (1980).

[3] Bose, N.K., Applied Multidimensional Systems Theory, van Nostrand Reinhold Comp., New York (1982).

[4] Bose, N.K. and J.P. Guiver, Multivariate polynomial positivity invariance under coefficient perturbation, IEEE Trans. Acoust. Speech Signal Process ASSP - 28, 660-665 (1980).

[5] Cargo, G.T. and O Shisha, The Bernstein form of a polynomial, J. Res. Nat. Bur. Standards 70B, 79-81 (1966).

[6] Grassmann, E. and J. Rokne, The range of values of a circular complex polynomial over a circular complex interval, Computing 23, 139-169 (1979).

[7] Hardy, G.H., J.E. Littlewood, and G. Pólya, Inequalities, 2nd ed., Cambridge Univ. Press, London and New York (1959).

[8] Lane, J.M. and R.F. Riesenfeld, Bounds on a polynomial, BIT 21, 112-117 (1981).

[9] Lorentz, G.G., Bernstein Polynomials, Univ. Toronto Press, Toronto (1953)

[10] Ratschek, H. and J. Rokne, Computer Methods for the Range of Functions, Ellis Horwood Ltd., Chichester (1984).

[11] Rheinboldt, W.C., C.K. Mesztenyi, and J.M. Fitzgerald, On the evaluation of multivariate polynomials and their derivatives, BIT 17, 437-450 (1977).

[12] Rivlin, T.J., Bounds on a polynomial, J. Res. Nat. Bur. Standards 74B, 47-54 (1970).

[13] Rokne, J., Bounds for an interval polynomial, Computing 18, 225-240 (1977).

[14] Rokne, J., A note on the Bernstein algorithm for bounds for interval polynomials, Computing 21, 159-170 (1979).

[15] Rokne, J., The range of values of a complex polynomial over a complex interval, Computing 22, 153-169 (1979).

[16] Rokne, J., Optimal computation of the Bernstein algorithm for the bound of an interval polynomial, Computing 28, 239-246 (1982).

[17] Walach, E. and E. Zeheb, Sign test of multivariable real polynomials, IEEE Trans. Circuits and Systems CAS-27, 619-625 (1980).

# ON AN INTERVAL COMPUTATIONAL METHOD FOR FINDING THE REACHABLE SET IN TIME-OPTIMAL CONTROL PROBLEMS

*Tadeusz GIEC*

*Institute of Mathematics*
*Łódź University, Poland*

## 1. INTRODUCTION

Consider the time-optimal control of a system described by the differential equation

$$\dot{x}(t) = A(t)x(t) + B(t)u(t) \tag{1.1}$$

with fixed initial data $x(t_o) = x_o$, where $x$ is the n-dimensional state vector, $u$ is the r-dimensional vector control function, $A(t)$, $B(t)$ are the $n\times n$, $n\times r$ - matrix functions, respectively. $A(t)$, $B(t)$, are assumed to be piecewise continuous for $t \geqq t_o$ on any finite interval.

A control function $u(t)$ is said to be admissible if it is measurable over any finite interval and takes its values from a given compact set $U$ of $E^r$.

Let $U$ denote the set of all admissible control functions. Given an initial state

$$x(t_o) = x_o \tag{1.2}$$

and a control function $u(t) \in U$, $t_o \leqq t \leqq t_1$, then equation (1.1) has a unique solution $x(t,u)$.

Let $\Phi(t)$ be the principle matrix solution of the homogeneous system

$$\dot{x}(t) = A(t)x(t) \tag{1.3}$$

satisfying $\Phi(t_o) = J$ - the identity matrix. Then, for an admissible control $u$, the solution of (1.1) is given by

$$x(t,u) = \Phi(t)\Phi^{-1}(t_o)x_o + \Phi(t) \int_{t_o}^{t} \Phi^{-1}(\tau)B(\tau)u(\tau)d\tau. \tag{1.4}$$

Applying the change of coordinates in equation (1.1), defined by the transformation $x = \Phi(t)y$, we get that system (1.1) is equivalent to

$$(1.5) \qquad \dot{y} = \psi(t)u(t)$$

with the initial data $y(t_o,u) = 0$, where $\psi(t) = \Phi^{-1}(t)B(t)$.

Define

$$(1.6) \qquad R(t) = \{y(t,u);\ u \text{ measurable},\ u(\tau) \in U \text{ for } \tau \in [t_o,t]\}$$

where

$$y(t,u) = \int_{t_o}^{t} \psi(\tau)u(\tau)d\tau.$$

$R(t)$ is called a reachable set.

The time-optimal control problem is to find an admissible control $u^*$, subject to its constraints, in such a way that the solution $x(t,u^*)$ of (1.1) reaches a continuously moving target in $E^n$ in minimum time $t^* \geq t_o$.

The equivalent statement of the time-optimal control problem is to find an admissible control $u$ for which $w(t) \in R(t)$ for a minimum value of $t \geq t_o$, where $w(t)$ stands for a moving target at time $t$.

## 2. PROPERTIES OF THE REACHABLE SET

We are now restricting ourselves to values of the control function in the unit cube $C^r$ of $E^r$. The set of admissible controls on $[t_o,t]$ is given by

$$(2.1) \qquad \Omega[t_o,t] = \{u:\ u \text{ measurable on } [t_o,t],\ u(\tau) \in C^r,\ t_o \leq \tau \leq t\}.$$

The reachable set $R(t)$ is then of the form

$$(2.2) \qquad R(t) = \{\int_{t_o}^{t} \psi(\tau)u(\tau)d\tau;\ u \in \Omega[t_o,t]\}.$$

It is known [2] that if $\Psi$ is an $n\times r$-matrix-valued function with components $\psi_{ij}$ in $L_1[t_o,t^*]$ and $\Omega$ is the set of r-vector-valued measurable functions $u$ whose components $u_j$ satisfy $|u_j(t)| \leq 1$, $j = 1,2,\dots,r$, and $\Omega^o$ is that subset of $\Omega$ for which $|u_j(t)| = 1$, then

$$\{\int_{t_o}^{t^*} \Psi(\tau)u(\tau)d\tau,\ u \in \Omega\} \text{ is convex, compact, and}$$

$$(2.3)\qquad \{\int_{t_o}^{t} \psi(\tau)u(\tau)d\tau;\quad u \in \Omega\} = \{\int_{t_o}^{t} \psi(\tau)u^o(\tau)d\tau;\quad u^o \in \Omega^o\}.$$

From this statement we have

THEOREM 1. The reachable set $R(t)$ is convex and compact.

Formula (2.3) contains the statement which is called in control theory the "bang-bang" principle. The set of bang-bang controls on $[t_o,t]$ is

$$(2.4)\qquad \Omega^o[t_o,t] = \{u;\ u \text{ measurable},\ |u_j(\tau)| = 1,\ j = 1,\dots,r,\ \tau \in [t_o,t]\}.$$

These are the controls which at all times utilize all the controls available. Then

$$(2.5)\qquad R^o(t) = \{\int_{t_o}^{t} \psi(\tau)u^o(\tau)d\tau,\ u^o \in \Omega^o[t_o,t]\}$$

is the set of points reachable by the bang-bang control.

THEOREM The Bang-Bang Principle :

$$(2.6)\qquad R(t) = R^o(t) \qquad \text{for each} \quad t \geq t_o.$$

The bang-bang principle says that any point that can be reached by an admissible control in time $t$ can also be reached by a bang-bang control in the same time

There are several other properties of the reachable set.

THEOREM 2. $R(t)$ is a continuous function on $[t_o,\infty)$.

P r o o f. Since, for each $t_o \geq 0$ and $t \geq t_o$, we have

$$|y(t,u) - y(t_o,u)| = |\int_{t_o}^{t} \psi(\tau)u(\tau)d\tau| \leq |\int_{t_o}^{t} \|\psi(\tau)\|d\tau|,$$

therefore, by the definitions of the reachable set and of the metric space,

$$\rho(R(t),R(t_o)) \leq |\int_{t_o}^{t} \|\psi(\tau)\|d\tau.$$

Since $\int_{t_o}^{t} \|\psi(\tau)\|d\tau$ is absolutely continuous, the theorem is true.

THEOREM 3. If $y$ is an interior point of $R(t^*)$ for some $t^*>t_o$, then it is an interior point of $R(t)$ for some $t \in (t_o,t^*)$.

P r o o f. Let $V$ be a neighbourhood of $y$ of radius $\delta$ inside $R(t^*)$. Suppose, for each $t \in (t_o,t^*)$, that $y$ is not an interior point of $R(t)$. Then there is a support hyperplane $p(\eta)$ through $y$, such that $R(t)$ lies on one side of $p(\eta)$. Let the neighbourhood $V$ of $y$ be inside $R(t^*)$; then there is a point $q$ of $R(t^*)$ whose distance from $R(t)$ is at least $\delta$ for each $t \in (t_o,t^*)$. This contradicts the continuity of $R(t)$ and completes the proof.

Let a nonzero vector $\eta$ define a direction in $E^n$. Suppose that we want to find an admissible control $u$ that maximizes the rate of change of $y(t,u)$ in the direction $\eta$, that is, we want to maximize

$$(2.7) \qquad \eta'\dot{y} = \eta'\psi(t)u(t).$$

We see that if $u^*$ is of the form

$$(2.8) \qquad u^*(t) = \text{sgn}\,[\eta'\psi(t)], \qquad \eta \neq 0,$$

then

$$\eta'\psi(t)u^*(t) = \sum_{j=1}^{r} |[\eta'\psi(t)]_j|.$$

Equation (2.8) means that, for each $j = 1,\dots,r$,

$$u_j^*(t) = \text{sgn}\,[\eta'\psi(t)]_j \quad \text{when} \quad [\eta'\psi(t)]_j \neq 0.$$

When $|u_j^*(t)| = 1$ almost everywhere for $j = 1,\dots,r$, we say that the control $u^*$ is bang-bang.

Thus, the control $u^*$ maximizes $\eta'y(t,u)$ over all admissible controls if and only if it is of form (2.8). So, for any fixed $t^* > t_o$ and any $u^*$ of form (2.8), the point $q^* = y(t^*,u^*)$ lies on the boundary of $R(t^*)$. Moreover, $\eta'(s - q^*) \leq 0$ for all $s \in R(t^*)$ and the hyperplane $p(\eta)$ through $q^*$ normal to $\eta$ is a support plane of $R(t^*)$ at $q^*$. Note also that if $u$ is any other control of form (2.8), then $y(t^*,u)$ lies on this hyperplane $p(\eta)$. Conversely, if $q^*$ lies on the boundary of $R(t^*)$, then there is a support plane $p(\eta)$ of $R(t^*)$ through $q^*$ and we may take $\eta$, which is a nonzero vector, being an outward normal. Hence we have proved the following

THEOREM 4. A point $q^* = y(t^*,u^*)$ is a boundary point of $R(t^*)$ with $\eta$ an outward normal to a support plane of $R(t^*)$ through $q^*$ if and only if $u^*$ is of the form $u^*(t) = \text{sgn}\,[\eta'\psi(t)]$ on $[t_o,t^*]$

for some $\eta \neq 0$.

Define

$$E_j(\eta) = \{t: \eta' y_j(t) = 0, \quad t \in [t_o, t^*]\}.$$

We say that system (1.1) is normal on $[t_o, t^*]$ if $E_j(\eta)$ has measure zero for each $j = 1, \ldots, r$ and each $\eta \neq 0$.

It is known [2] that system (1.1) is normal on $[t_o, t^*]$ if and only if $R(t^*)$ is strictly convex.

## 3. A COMPUTATIONAL METHOD

As it follows from the considerations above, in the computational realization we have to compute the minimum time $t^* = \inf\ \{t, \ p \in R(t)\}$ and to find the reachable set $R(t)$ at time $t^*$.

We now want to present an algorithm for finding the minimum time $t^*$.

Let $\alpha(\eta, t)$ be a function defined by

$$(3.1) \qquad \alpha(\eta, t) = \eta' \beta(t) - \eta' w(t) \quad \text{for} \quad t \geq t_o, \qquad \eta \in E^n,$$

where

$$\eta' \beta(t) = \max_{y \in R(t)} \eta' y.$$

It can easily be seen that a necessary and sufficient condition for $w(t')$ to belong to $R(t')$ for some $t'$ is

$$\alpha(\eta, t') \geq 0 \quad \text{for each} \quad \eta \in E^n.$$

The function $\alpha(\eta, t)$ is continuous, hence the set

$$(3.2) \qquad T = \{t: \alpha(\eta, t) < 0\}$$

is open for any $\eta \in E^n$ and can be composed as follows:

$$(3.3) \qquad T = \bigcup_{n=1}^{\infty} (a_n, b_n).$$

Let $\tau \geq t_o$, and let $N_\tau$ denote the set of indices $m$ for which

$$a_m < \tau, \qquad b_m > t_o \quad \text{if} \quad \tau > t_o,$$

and

$$a_m \leq \tau, \qquad b_m > t_o \quad \text{if} \quad \tau = t_o.$$

Let $p(\eta,\tau)$ be defined for $\eta \in S$ ($S$ - the unit sphere of $E^n$) and $\tau \geq t_o$ as

$$(3.4) \qquad p(\eta,\tau) = \begin{cases} \sup\limits_{m \in N} b_m & \text{if } N_\tau \neq \emptyset \\ \tau - \alpha(\eta,\tau) & \text{if } N_\tau = \emptyset. \end{cases}$$

If it is known that $w(t) \notin R(t)$, $t_o \leq t < \tau_o$ and, for some $\eta \in S$, $\tau_1 = p(\eta,\tau_o) > \tau_o$, then $w(t) \notin R(t)$, $t_o \leq t < \tau_1$.

Let $(\eta_k)$ $(k = 0,1,\ldots,\eta_k \in S)$ and $t_o$ be given.

Assign the sequence $(\tau_k)$ to them as

$$\begin{array}{ll} \tau_o = t_o, & t_1 = p(\eta_o,\tau_o), \\ \tau_1 = \max\{t_1,\tau_o\}, & t_2 = p(\eta_1,\tau_1), \\ \cdots\cdots\cdots & \cdots\cdots\cdots \\ \tau_n = \max\{t_n,\tau_{n-1}\}, & t_{n+1} = p(\eta_n,\tau_n). \end{array}$$

The sequence $(\tau_k)$ is nondecreasing and bounded with respect to $(\eta_i)$ as a consequence of the assumption that there exists a solution of problem (1.1). Therefore $\lim\limits_{i\to\infty} \tau_k$ exists.

DEFINITION 1. The sequence $(\eta_k)$ is said to be maximizing if, for the corresponding sequence $(\tau_i^*)$, the equality

$$(3.5) \qquad \tau^* = \lim_{k\to\infty} \tau_k^* = \sup_{(\eta_j)\in S} (\lim_{k\to\infty} \tau_k) = \tau^o$$

holds.

It is easily proved that there exists a maximizing sequence.

THEOREM 5. Let $\tau^*$ be determined by a maximizing sequence. Then $\tau^* = t^*$ is the optimal time of the control problem.

P r o o f. Suppose the contrary. Then only the inequality $\tau^* < t^*$ can hold, hence $w(t) \notin R(t)$ for $t \in [t_o,\tau^*]$. Thus there exists an $\eta^o$ for which $-\varepsilon = \alpha(\eta^o,\tau^*) < 0$ with some positive $\varepsilon$. Since $\alpha(\eta,t)$ is continuous, there exists a neighbourhood $V$ of $(\eta^o,\tau^*)$ such that, for any $(\eta,\tau) \in V$, we have

$$|\alpha(\eta,\tau) - \alpha(\eta^o,\tau^*)| < \frac{\varepsilon}{2}$$

and

$$\alpha(\eta,\tau) < -\frac{\varepsilon}{2}.$$

Moreover, there exists an $n_o$ such that $\tau_n^* \in (\tau',\tau")$ for $n \geq n_o$. Consider any $\bar{\eta} \in V_{\eta_o}$ and let

$$\tilde{\eta}_k = \begin{cases} \eta_k^*, & k < n_o \\ \bar{\eta}, & k \geq n_o. \end{cases}$$

From $p(\bar{\eta},\tau_{n_o}) > \tau" > \tau^* = \tau^o$ it follows that $\lim_{k\to\infty} \tilde{\tau}_k > \tau^o$ for $(\tilde{\eta}_k)$, which contradicts the definition of $\tau^o$.

It is very difficult to compute the set $R(t)$. For the purpose, using formula (2.2), we have to know the optimal control $u^*$ with all its switching points. In the general case, it is not possible to find exactly all the switching points. In order to avoid these difficulties, we shall use the interval integrals.

Let $F$ denote a function on $[0,\infty)$ to the vector interval space. Using the definition of the interval integral, we define the vector interval integral.

DEFINITION 2. If $F = (f_1,\ldots,f_n)$ is a vector interval function defined on the interval $[a,b]$, then the vector integral of $F$ over $[a,b]$ is defined to be the interval vector

$$\int_a^b F(t)dt = \left(\int_a^b f_1(t)dt,\ldots,\int_a^b f_n(t)dt\right) \tag{3.6}$$

where $\int_a^b f_i(t)dt$, $i = 1,\ldots,n$, denote the interval integrals.

Define

$$[\varphi_{ij}(t)]_{n\times r} = \Phi^{-1}(t)B(t). \tag{3.7}$$

THEOREM 6. Let us assume that system (1.1) is normal; then, for any nonzero vector $\eta$, there exists an optimal control $u^*$ of the form

$$u^*(t) = \mathrm{sgn}\,[\eta'\Phi^{-1}(t)B(t)] \tag{3.8}$$

for which the reachable set at time $t^*$ satisfies the best possible inclusion

$$(3.9)\qquad R(t^*) \subset \left(\left[-\int_{t_o}^{t^*} f_1(\tau)d\tau, \int_{t_o}^{t^*} f_1(\tau)d\tau\right],\dots,\left[-\int_{t_o}^{t^*} f_n(\tau)d\tau, \int_{t_o}^{t^*} f_n(\tau)d\tau\right]\right)$$

where $\Phi(t)$ is the principle matrix solution of homogeneous system (1.3), $\Phi^{-1}(t)B(t)Q = [\varphi_{ij}(t)]_{n\times r}Q$ stands for an extended vector interval function, $Q$ is the r-dimensional interval vector whose coordinates are the intervals of the form $[-1,1]$,

$$f_i(t) = \varphi_{i1}(t) + \dots + \varphi_{ir}(t), \qquad i = 1,2,\dots,n.$$

P r o o f. The result follows from a series of simple observations. Since system (1.1) is normal, the boundary points of $R(t^*)$ can, and can only, be reached by a control that is bang-bang. By theorem 4, there exists an optimal control of form (3.9). Thus, the optimal control is the r-dimensional vector function whose coordinates take the values $-1$ or $1$. Consequently, the optimal control will be contained in the interval vector $Q$ whose coordinates are the intervals of the form $[-1,1]$. Multiplying the matrix $\Phi^{-1}(t)B(t)$ by the interval vector $Q$, we get the vector interval function. Applying the methods for integration of interval functions and the Bang-Bang Principle, we obtain formula (3.9). On the other hand, since system (1.1) is normal, it follows that the reachable set $R(t^*)$ is strictly convex, so the best possible inclusion (3.9) holds.

EXAMPLE. Consider a simple control system $\dot{x}_1 = u_1$, $\dot{x}_2 = u_2$, $|u_1| \leq 1$, $|u_2| \leq 1$. Here

$$B = \begin{pmatrix} 1 & 0 \\ 0 & 1 \end{pmatrix}.$$

The vector interval function will be of the form

$$\Phi^{-1}(t)B(t)Q = \begin{pmatrix} 1 & 0 \\ 0 & 1 \end{pmatrix} \begin{pmatrix} [-1,1] \\ [-1,1] \end{pmatrix}.$$

The reachable set $R(t^*)$ is given by the formula

$$R(t^*) = \begin{pmatrix} [-t^*,t^*] \\ [-t^*,t^*] \end{pmatrix}.$$

This means that $R(t^*)$ is a square with sides of length $2t^*$.

REFERENCES

[1] T. Giec, Application of Interval Analysis to Linear Problems of Optimal Control with Quadratic Cost Functional, Interval Mathematics (1980).

[2] H. Hermes, J.P. LaSalle, Functional Analysis and Time Optimal Control, Academic Press, New York and London (1969).

[3] R.E. Moore, Methods and Application of Interval Analysis, SIAM Publications, Philadelphia, Pa., (1979).

[4] K. Nickel, Schranken für die Lösungsmengen von Funktional-Differential - Gleichungen, Freiburger Intervall - Berichte 79/4.

[5] E. Nuding, Computing the Exponential of an Essentially Nonnegative Matrix, Interval Mathematics (1980).

[6] L. Rall, Numerical Integration and the Solution of Integral Equation by the Use of Riemann Sums, SIAM Review 7 (1965), 55-64.

[7] E.Gyurkovics, T. Vörös, On a Computational Method of the Time--Optimal Control Problem, Colloquia Methematica Societatis Janos Bolyai (1977).

[8] S. Walczak, A Note on the Controllability of Nonlinear Systems, Mathematical Systems Theory 17, (1984), 351-356.

# ON THE OPTIMALITY OF INCLUSION ALGORITHMS

Henryk Kołacz

Institute of Mathematics

Technical University of Poznań

Poznań, Poland

Abstract. In this paper a general concept of inclusion algorithm is introduced. Any inclusion algorithm provides a set that includes the solution of a given problem. Inclusion algorithms are studied with respect to the information used by them.
Some examples illustrate the presented concepts and results.

## 1. Introduction

In computational practice we must take into consideration that the rounding and propagated errors can give a large and inestimable error of the final result.
In general, it is difficult to provide a priori estimates of this error, and even in case they are available they produce bounds so pessimistic that they are of little practical importance.
Therefore there is a need for automatic error control in numerical computations.
A very useful tool for it is the interval analysis introduced by Moore [2]. The basic idea of this analysis is the inclusion of the solution of a given problem by intervals.
In this paper we introduce the concept of inclusion algorithm. It is defined as an arbitrary operator $\varphi$ such that it provides a set including the solution of a given problem. We shall assume that there exists an arithmetic such that the computed values of $\varphi$ are outer approximations of the exact values of $\varphi$ .
We present a model of optimality for inclusion algorithms. It is based on the methodology introduced by Traub and Woźniakowski in [7].
The optimality of inclusion algorithms is studied with respect to error and computational complexity. It is shown that the intersection algorithm is a strongly optimal inclusion algorithm with respect to error. There are some connections between our optimality model and the ideas of Ratschek [6].
To illustrate concepts and results we present two examples: integration and range approximation.

## 2. Basic definitions

Let $E, F$ be two given sets. By $\mathbb{P}(E)$ we denote the power set of $E$, that is, the class of all subsets of $E$.
Let $R_E \subset \mathbb{P}(E)$ be a fixed class of subsets of $E$. The family $R_E$ is called a class of set representations in $E$.
For example $R_E$ is the class of all closed balls in a pseudometric space $E$ or the class of all closed intervals in an ordered space $E$.
We assume that there exists an operator $H: \mathbb{P}(E) \rightarrow R_E$ such that:

(2.1) $\qquad H(X) = X \quad$ for all $X \in R_E$ ,

(2.2) $\qquad X \subset H(X) \quad$ for all $X \in \mathbb{P}(E)$,

(2.3) $\qquad X \subset Y$ implies $H(X) \subset H(Y)$ for all $X, Y \in \mathbb{P}(E)$.

The operator $H$ satisfying the properties (2.1)-(2.3) is called a monotone upwardly directed rounding (see [2]).

In our model we assume that the distance between elements of the family $R_E$ is measured by elements of a complete lattice $K$. Then every subset of $K$ has an infimum and a supremum.
Moreover, let $\inf K = \theta$, that is, $m \geq \theta$ for all $m \in K$.

Definition 2.1. We shall say that $d\colon R_E \times R_E \to K$ is a distance operator in the class $R_E$ if

(2.4) $\qquad X, Y \subset Z \Rightarrow d(X,Y) \leq d(X,Z),\ d(Y,Z)$

for all $X, Y, Z \in R_E$ .

Let $\varepsilon$ be a given element of $K$, $\varepsilon \geq \theta$.

Definition 2.2. We shall say that $X \in R_E$ is an $\varepsilon$-inclusion of an element $x \in E$ if

$1^\circ$ $\quad x \in X$,

$2^\circ$ $\quad d(H(x), X) \leq \varepsilon$ .

We illustrate the above concepts by an example.

Example 2.1. Let $E$ be a normed linear space over the real or complex field. Let $R_E$ be an arbitrary class of set representations in $E$ such that it includes the class of all singletons in $E$.
We define the distance operator $d$ in $R_E$ as

$$d(X,Y) = \|X - Y\| ,$$

where $\|X\| = \sup[\|x\|\colon x \in X]$. The set $U(x,\varepsilon)$ defined as

$$U(x,\varepsilon) = \{X \in R_E\colon x \in X,\ \|x - X\| \leq \varepsilon\},$$

is the family of all $\varepsilon$-inclusions of an element $x \in E$, where $\varepsilon$ is a fixed nonnegative real number.

## 3. Information operators

Let $S: F \to E$ be an arbitrary operator. We want for any $f \in F$ to find an $\varepsilon$-inclusion of $S(f)$. To find it, we must know something about the element f. Let

$$N: F \to \mathcal{H} \tag{3.1}$$

be an arbitrary operator, where $\mathcal{H}$ is a given space.
The operator N is called the basic information operator for F and the element N(f) is called the basic information of f.

Definition 3.1. Let $f \in F$ and $\mathcal{U}$ be a given set. We shall say that $L: f \to \mathcal{U}$ is an information operator for f (generated by N) if $N(f) \subset L(f)$.

We denote the family of all information operators for f, $f \in F$ by $\hat{I}_N(f)$. Obviously $\hat{I}_N(f)$ is nonempty for all $f \in F$ because $N \in \hat{I}_N(f)$. We illustrate the concept of information operator by the following example.

Example 3.1. Let M be a Banach space over the field of real numbers $\mathbb{R}$ and A be a nonempty subset of M.
Let $\mathcal{F}$ be a nonempty class of operators mapping A into M, which are n-times Frechet differentiable on A, where n denotes a fixed natural number. We take $F = \mathcal{F} \times R_A$ and $\mathcal{H} = R_M \times R_M \times \ldots \times R_M$ ((n+1)-times), where $R_A$ and $R_M$ denote fixed classes of set representations in A and M, respectively. Let $H: \mathbb{P}(M) \to R_M$ be a monotone upwardly directed rounding.

We define the basic information operator N in the following way:

$$N(g,X) = \left[ H(\bar{g}(X)), H(\bar{g}'(X)), \ldots, H(\bar{g}^{(n)}(X)) \right],$$

where $\bar{g}^{(j)}(X)$ denotes the range of the jth Frechet derivative of $g \in \mathcal{F}$ over X. Then every information operator $L_g \in \hat{I}_N(g)$ has the following form:

$$L_g(X) = \left[ G(X), G'(X), \ldots, G^{(n)}(X) \right],$$

where $\mathcal{H} = \mathcal{U}$ and $G^{(j)}$ is an extension of $g^{(j)}$ i.e. $\bar{g}^{(j)}(X) \subset$ $\subset G^{(j)}(X)$ for all $X \in R_A$ and $j = 0,1,2,\ldots,n$.

The inclusion between elements of the space $\mathcal{H}$ is meant componentwise.

It is often necessary to impose some restrictions on $L \in \hat{I}_N(f)$ in order to guarantee that the information $L(f)$ can be easily computed and enjoys some useful properties.
Let $I_N$ be an operator defined on the set $F$ such that $I_N(f)$ is a given family of information operators for $f \in F$, $I_N(f) \subset \hat{I}_N(f)$.
The operator $I_N$ is called an information selection operator for $F$.
We denote

$$(3.2) \qquad I_N(F) = \{ N_f : N_f \in I_N(f), \quad f \in F . \}.$$

Example 3.2. Let $\mathcal{H}$ and $\mathcal{U}$ be given nonempty families of subsets of a space $T$. Let $d$ be a distance operator in $P(T)$ with values in $C = [0,+\infty)$ and be a fixed nonnegative real number. Then the operator $I_N$ defined as

$$I_N(f) = \{ L \in \hat{I}_N(f) : \ d(N(f), L(f)) \leq \varepsilon \},$$

is an information selection operator for $F$.

For a given element $f \in F$ and an information operator $L \in I_N(f)$ we define the set $V(f,L)$ as follows:

$$(3.3) \qquad V(f,L) = \{ g \in F: \ \text{there exists } M \in I_N(g) \text{ such that } L(f)=M(g) \}.$$

Therefore $V(f,L)$ is the set of all elements $g \in F$ which have the same information as $f$ under $L$. It is obvious that $V(f,L)$ is nonempty for every $f \in F$, $L \in I_N(f)$ because $f \in V(f,L)$.
Knowing $L(f)$, it is impossible to recognize which element $S(f)$ or $S(g)$ is being actually approximated for all $g \in V(f,L)$.
Analogously as in [7] we introduce the following definition.

Definition 3.2. We shall say din $(I_N,f)$ is the local diameter of information if

$$(3.4) \qquad \mathrm{din}(I_N,f) = \sup_{L \in I_N(f)} \ \sup_{g_1,g_2 \in V(f,L)} d(H(S(g_1)),H(S(g_2))).$$

We shall say $\mathrm{din}(I_N)$ is the (global) diameter of information if

(3.5) $$\operatorname{din}(I_N) = \sup_{f\in F} \operatorname{din}(I_N, f).$$

4. Error of inclusion algorithms

To determine an $\varepsilon$-inclusion of $S(f)$ we use an inclusion algorithm which is an operator defined as follows.

Definition 4.1. We shall say that $\varphi : I_N(F) \to R_E$ is an inclusion algorithm for the problem $S$ if

(4.1) $$S(f) \in \varphi(N_f)$$

for all $f \in F$ and $N_f \in I_N(f)$.

We denote the class of all inclusion algorithms using the information generated by the information selection operator $I_N$ by $\hat{A}(I_N)$.

Let us observe that $\hat{A}(I_N)$ is an ordered set with the order relation $\leqslant$ defined as follows:

(4.2) $$\varphi_1 \leqslant \varphi_2 \Longleftrightarrow \varphi_1(N_f) \subset \varphi_2(N_f)$$

for all $f \in F$ and $N_f \in I_N(f)$, where $\varphi_1, \varphi_2 \in \hat{A}(I_N)$.

Definition 4.2. We shall say $e(\varphi, f)$ is the local error of $\varphi \in \hat{A}(I_N)$ if

(4.3) $$e(\varphi, f) = \sup_{N_f \in I_N(f)} \ \sup_{g \in V(f, N_f)} d(H(S(g)), \varphi(N_f)).$$

We shall say $e(\varphi)$ is the (global) error of $\varphi$ if

(4.4) $$e(\varphi) = \sup_{f\in F} e(\varphi, f).$$

It is obvious that if $\varphi_1 \leqslant \varphi_2$ then $e(\varphi_1) \leqslant e(\varphi_2)$ for all inclusion algorithms $\varphi_1, \varphi_2 \in \hat{A}(I_N)$.

From the inclusion (4.1) it follows that the local diameter of information is a lower bound on the local error of any inclusion algorithm. A formal proof is provided by

Theorem 4.1. For any inclusion algorithm $\Phi \in \hat{A}(I_N)$,

(4.5) $$e(\Phi, f) \geqslant \mathrm{din}(I_N, f)$$

for all $f \in F$. Moreover,

(4.6) $$e(\Phi) \geqslant \mathrm{din}(I_N).$$

Proof. Let $f \in F$ and $N_f \in I_N(f)$. It is obvious that $S(g) \in \Phi(N_f)$ for all $g \in V(f, N_f)$. From this by the formula (2.4) we obtain the inequality (4.5).
The inequality (4.6) is a simple consequence of (4.5). The proof is complete.

Example 4.1. For $f \in F$, $N_f \in I_N(f)$ we define

(4.7) $$U^*(N_f) = H(\{S(g): \; g \in V(f, N_f)\}).$$

It is obvious that $U^*$ is an inclusion algorithm, $U^* \in \hat{A}(I_N)$.
From the inclusion (4.1) it follows that

(4.8) $$U^*(N_f) \subset \Phi(N_f),$$

for all $f \in F$, $N_f \in I_N(f)$ and any inclusion algorithm $\Phi$.
Moreover, taking $R_E := \mathbb{P}(E)$ and $d(X,Y) = \|X-Y\|$ we obtain

(4.9) $$e(U^*, f) = \mathrm{din}(I_N, f)$$

for all $f \in F$. This means that the inequalities (4.5), (4.6) cannot be improved in general.

Let $A(I_N)$ be a nonempty class of inclusion algorithms using the information generated by $I_N$.

Definition 4.3. We shall say that $P \in A(I_N)$ is a strongly optimal error inclusion algorithm in the class $A(I_N)$ if

(4.10) $$\inf\left[e(\Phi, f): \Phi \in A(I_N)\right] = e(P, f)$$

for all $f \in F$.
We shall say that $P \in A(I_N)$ is an optimal error inclusion algorithm in the class $A(I_N)$ if

(4.11) $$\inf\left[e(\Phi): \Phi \in A(I_N)\right] = e(P).$$

Theorem 4.2. Let $A(I_N)$ be a nonempty family of inclusion algorithms such that $A(I_N) = \hat{A}(I_N) \cap W$, where $W$ is a class of set operators. We define the operator $\varphi^*$ as

$$\varphi^*(N_f) = H\Big(\bigcap_{\varphi \in A(I_N)} \varphi(N_f)\Big) . \tag{4.12}$$

Suppose $\varphi^* \in W$. Then $\varphi^*$ is a strongly optimal error inclusion algorithm in $A(I_N)$.

Proof. First let us observe that $\varphi^*$ is an inclusion algorithm. Therefore $\varphi^* \in A(I_N)$. Since $\varphi^* \leq \varphi$ for any inclusion algorithm $\varphi \in A(I_N)$, $e(\varphi^*,f) \leqslant e(\varphi,f)$ for all $f \in F$. From this we obtain that $\varphi^*$ is a strongly optimal error inclusion algorithm in $A(I_N)$. The proof is complete.

Corollary 4.1. The algorithm $U^*$ defined by the formula (4.7) is a strongly optimal error inclusion algorithm in the class $\hat{A}(I_N)$.

Proof. It is a simple consequence of the inclusion (4.8).

Remark 4.1. A strongly optimal error inclusion algorithm is also an optimal error inclusion algorithm but the converse is, in general, not true. Obviously $U^*$ is an optimal error inclusion algorithm in $\hat{A}(I_N)$.

## 5. Complexity of inclusion algorithms

In this section we present a model of computation which consists of a set of primitive operations, permissible information operators, and permissible inclusion algorithms. This model is based on the general setting given in [7].

(i) Let $t$ be a primitive operation in a given class of set representations $R_E$ in E. Examples of primitive operations in $I(E)$ are interval operations (the addition of two intervals, the multiplication of an interval by a real number etc.). Usually primitive operations in $R_E$ are defined by some corresponding operations in the space E (see [2]).

Let $T$ be a given set of primitive operations in $R_E$. We denote the complexity (the total cost) of $t$ by comp(t). We assume that comp(t) is finite.

(ii) Let $f \in F$ and $L \in I_N(f)$. We say that $L$ is a permissible information operator for $f$ with respect to $T$ if there exists a program using a finite number of primitive operations from $T$ which computes $L(f)$. We assume that if $L(f)$ requires the evaluation of operations $t_1, t_2, \ldots, t_k \in T$, then $\mathrm{comp}(L(f)) = \sum_{i=1}^{k} \mathrm{comp}(t_i)$.

(iii) Let $I_N(f)$ be a nonempty class of permissible information operators for $f$, $f \in F$. Let $\Phi \in \hat{A}(I_N)$. We say that $\Phi$ is a permissible inclusion algorithm with respect to $T$ if for every $f \in F$ and $L \in I_N(f)$ there exists a program using a finite number of primitive operations from $T$ which computes $Z \in R_E$ such that $Z \supset \Phi(Y)$, where $Y = L(f)$.
Let $\mathrm{comp}(\Phi(Y))$ be the complexity of computing $\Phi(Y)$. We assume that if $\Phi(Y)$ requires the evaluation of $s_1, s_2, \ldots, s_m \in T$, then $\mathrm{comp}(\Phi(Y)) = \sum_{i=1}^{m} \mathrm{comp}(s_i)$. We denote the class of all permissible inclusion algorithms with respect to $T$ in $\hat{A}(I_N)$ by $\hat{A}_T(I_N)$.
We define the complexity of $\Phi \in \hat{A}_T(I_N)$ as

$$(5.1) \qquad \mathrm{comp}(\Phi) = \sup_{f \in F} \sup_{L \in I_N(f)} \left[ \mathrm{comp}(L(f)) + \mathrm{comp}(\Phi(L(f))) \right].$$

Let $\mathcal{E} \geqslant \theta$ be a fixed element of a complete lattice $K$.
Let $A_T(I_N, \mathcal{E})$ be a nonempty subset of $\hat{A}_T(I_N)$ such that $e(\Phi) \leq \mathcal{E}$ for all $\Phi \in A_T(I_N, \mathcal{E})$.

<u>Definition</u> 5.1. We shall say that $P \in A_T(I_N, \mathcal{E})$ is an $\mathcal{E}$-complexity optimal inclusion algorithm in the class $A_T(I_N, \mathcal{E})$ if

$$(5.2) \qquad \inf \left[ \mathrm{comp}(\Phi) : \Phi \in A_T(I_N, \mathcal{E}) \right] = \mathrm{comp}(P).$$

The analysis needed to characterize and construct an $\mathcal{E}$-complexity optimal algorithm for a particular problem can be a difficult mathematical problem.

## 6. Applications

In this section we show some examples of how the above analysis can be applied to some concrete problems.

We present two examples: integration and range approsimation.

### (i) Integration

Let $F$ be the class of all continuous real functions defined on the interval $[a,b] \subset \mathbb{R}$. We take $E = \mathbb{R}$ and $R_E = I(\mathbb{R})$, where $I(\mathbb{R})$ denotes the class of all closed intervals over $\mathbb{R}$.

We define the distance operator $d$ in $I(\mathbb{R})$ as

$$d(X,Y) = \sup\left[\, |x-y| : x \in X,\ y \in Y \,\right]. \tag{6.1}$$

We define the operator $S\colon F \to \mathbb{R}$ as

$$S(g) = \int_a^b g(t)\,dt, \tag{6.2}$$

for $g \in F$.

Let $M$ be a positive integer and subdivide $[a,b]$ into $M$ subintervals $X_1, X_2, \dots, X_M$, so that

$$a = \underline{X}_1 < \overline{X}_1 = \underline{X}_2 < \overline{X}_2 < \dots < \overline{X}_M = b, \tag{6.3}$$

where $X_i = [\underline{X}_i, \overline{X}_i]$ for $i = 1,2,\dots,M$.

We define the basic information operator $N$ as

$$N(g) = \left[\bar{g}(X_1), \bar{g}(X_2), \dots, \bar{g}(X_M)\right], \tag{6.4}$$

where $g \in F$. Then any information operator for $g$ has the form:

$$L(g) = \left[\, G(X_1), G(X_2), \dots, G(X_M)\right], \tag{6.5}$$

where $G$ is an interval extension of $g$. The inclusion between elements of $I^M(\mathbb{R})$ (the Cartesian product of $I(\mathbb{R})$, M-times) is meant componentwise. For $g \in F$ and $L_g \in I_N(g)$ we define the interval operator as follows (see [3]):

$$\varphi(L_g) = \sum_{i=1}^{M} G(X_i)\,w(X_i), \tag{6.6}$$

where $w(X)$ denotes the width of an interval $X \in I(\mathbb{R})$.

Obviously by the mean value theorem $\varphi$ is an inclusion algorithm.

Let $Ex(g)$ be a nonempty family of interval extensions of $g \in F$.

Let $I_N(g)$ be the family of all information operators for $g$ of the form (6.5) with $G \in Ex(g)$.
Then it is not difficult to verify that

$$(6.7) \qquad din(I_N, g) = \sup_{G \in Ex(g)} \sum_{i=1}^{M} w(G(X_i))w(X_i).$$

Moreover, let us observe that

$$(6.8) \qquad e(\varphi, g) \leqslant \sup_{G \in Ex(g)} \sum_{i=1}^{M} w(G(X_i))w(X_i).$$

From this by Theorem 4.1 we obtain that $\varphi$ is a strongly optimal error inclusion algorithm.

(ii) Range approximation

Let $U$ be the family of all real functions defined on an interval $D \subset \mathbb{R}$ and differentiable $n$-times on $D \in I(\mathbb{R})$.
We take $F = U \times I(D)$, $E = I(\mathbb{R})$ and $R_E = I(\mathbb{R})$.
We define the distance operator $d$ in $I(\mathbb{R})$ by the formula (6.1).
For $X \in I(D)$ we define the power $X^n$ of $X$ by $X^n = \{x^n: \ x \in X\}$, where $n \geqslant 0$. We denote the absolute value of $X \in I(D)$ by $|X|$.
We define the operator $S: U \times I(D) \to I(\mathbb{R})$ as

$$(6.9) \qquad S(g,X) = \bar{g}(X),$$

where $\bar{g}(X)$ denotes the range of $g$ over $X$.
Let $N$ be the basic information operator defined as

$$(6.10) \qquad N(g,X) = \left[g(c), g'(c), \ldots, g^{(n-1)}(c), \bar{g}^{(n)}(X)\right],$$

where $c = m(X)$ is the midpoint of $X$, $n \in N$ and $g^{(j)}$ denotes the jth derivative of $g$. We define an information operator $N_g$ for $g$ as

$$(6.11) \qquad N_g(X) = \left[g(c), g'(c), \ldots, g^{(n-1)}(c), G^{(n)}(X)\right],$$

where $G^{(n)}$ is an interval extension of $g^{(n)}$.
Let $Ex(g^{(n)})$ be a nonempty class of interval extensions of $g^{(n)}$.
Let $I_N(g)$ be the family of information operators for $g$ of the form (6.11) with $G^{(n)} \in Ex(g^{(n)})$. We denote $w := w(D)$ and

$$(6.12) \qquad |E_n| := \sup_{G^{(n)} \in Ex(g^{(n)})} \left|G^{(n)}(D)\right|.$$

Developing functions with the same information as $g$ in Taylor series around $c$ we obtain

$$(6.13)\quad \mathrm{din}(I_N,g) \leqslant \sum_{k=1}^{n-1} 2^{\lambda_k - k} \frac{1}{k!} g^{(k)}(c)w^k + 2^{\lambda_n - n} \frac{1}{n!} |E_n| w^n ,$$

where

$$(6.14)\quad \lambda_k = \begin{cases} 0 & \text{if } k \text{ is even,} \\ 1 & \text{if } k \text{ is odd.} \end{cases}$$

For $g \in U$ and $X \in I(D)$ the Taylor form of $g$ of order $n$, is defined by (see [4]):

$$(6.15)\quad \Phi(N_g,X) = \sum_{k=0}^{n-1} \frac{1}{k!} g^{(k)}(c)(X-c)^k + \frac{1}{n!} G^{(n)}(X)(X-c)^n .$$

It is obvious that $\Phi$ is an inclusion algorithm for the problem $S$. It is not difficult to verify that

$$(6.16)\quad e(\Phi,g) \leqslant \sum_{k=1}^{n-1} 2^{\lambda_k - k} \frac{1}{k!} g^{(k)}(c)w^k + 2^{\lambda_n - n} \frac{1}{n!} |E_n| w^n .$$

Now let $U$ be the class of all polynomials of degree at most $n-1$ defined on the interval $D$. We take $\mathrm{Ex}(g^{(n)}) = \{G^{(n)}\}$, where $G^{(n)}(X) = [0,0]$ for all $X \in I(D)$.
Suppose $g^{(k)}(c) \geqslant 0$ or $g^{(k)}(c) \leqslant 0$ for $k = 1,2,\ldots,n-1$.
We shall present our consideration for the first assumption. The considerations for the second assumption are analogous.

$1^o$ Let $g^{(k)}(c) = 0$ for $k=2,4,6,\ldots$ . Then it is easily verified that

$$(6.17)\quad \mathrm{din}(I_N,g) = \sum_{k=1,\mathrm{odd}}^{n-1} 2^{1-k} \frac{1}{k!} g^{(k)}(c)w^k .$$

In this case the Taylor form has the following form:

$$(6.18)\quad \Phi(N_g,X) = g(c) + \sum_{k=1,\mathrm{odd}}^{n-1} \frac{1}{k!} g^{(k)}(c)[-z^k,z^k],$$

where $z = w(X)/2$. It is easy to show that

$$(6.19)\quad e(\Phi,g) \leqslant \mathrm{din}(I_N,g).$$

Therefore by Theorem 4.1, $\Phi$ is a strongly optimal error inclusion algorithm.

$2^{\circ}$ Let $g^{(k)}(c) = 0$ for $k = 1,3,5,\ldots$ . From this it follows that

$$\operatorname{din}(I_N,g) = \sum_{k=2,\text{even}}^{n-1} 2^{-k}\frac{1}{k!}\,g^{(k)}(c)w^k . \tag{6.20}$$

We have in this case

$$\Phi(N_g,X) = g(c) + \sum_{k=2,\text{even}}^{n-1} \frac{1}{k!}\,g^{(k)}(c)\,[0,z^k] . \tag{6.21}$$

It is easy to verify that in this case the inequality (6.19) holds, too. Therefore $\Phi$ is a strongly optimal error inclusion algorithm.

The problems connected with the range approximation by Taylor forms were considered in [3],[4],[5] (see also bibliography in [4]).

## References

[1] Kulisch, U.W. and Miranker, W.L.: Computer arithmetic in theory and practice, Academic Press, New York, 1981.

[2] Moore, R.E.: Interval analysis, Printice-Hall, Englewood Cliffs, New York, 1966.

[3] Moore, R.E.: Methods and applications of interval analysis, SIAM, Philadelphia, 1979.

[4] Ratschek, H. and Rokne, J.: Computer methods for the range of functions, Ellis Horwood Limited, 1984.

[5] Ratschek, H.: Optimality of the centered form for polynomials, Journal of Approximation Theory, 32, pp. 151-159, 1981.

[6] Ratschek, H.: Optimal approximations in interval analysis, in: Interval Mathematics, ed. K. Nickel, Academic Press, pp. 181-202, 1980.

[7] Traub, J.F. and Woźniakowski, H.: A general theory of optimal algorithms, Academic Press, New York, 1980.

INTERVAL OPERATORS AND FIXED INTERVALS

by R. Krawczyk

## 1. Introduction

In order to enclose a solution $x^*$ of a nonlinear system of equations $g(x) = 0$, where $g\colon B \subseteq \mathbb{R}^n \to \mathbb{R}^n$, many interval operators $F\colon \mathbb{I}B \to \mathbb{I}\mathbb{R}^n$ with the property

$$x^* \in X \;\Rightarrow\; x^* \in F(X) \qquad (*)$$

are discussed.

By applying the iteration method

$$X_0 \subseteq B,\; X_{k+1} := F(X_k),\; k = 0,1,2,\ldots$$

we obtain a monotone sequence of intervals

$$X_0 \supseteq X_1 \supseteq X_2 \supseteq \ldots$$

We distinguish between two cases:

1. There exists a $k \in \mathbb{N}$ with $X_{k+1} = \emptyset$. Then, because of (*) $X_0$ contains no solution.

2. The sequence $\{X_k\}$ is infinite. It then follows that $\lim_{k\to\infty} X_k = X_\infty$.

2.1 Additionally, if $\operatorname{rad} X_\infty = 0$ then $X_\infty = x^*$ is a unique solution.

2.2 On the other hand, if a solution $x^* \in X_0$ exists then it follows that $x^* \in X_k$ for all $k \in \mathbb{N}$ and $x^* \in X_\infty$.

In all cases, we assume that the Jacobi matrix $g'$ of $g$ exists, and that we know an interval extension $G'$ of $g'$, or more generally, that $g$ fulfills an interval Lipschitz condition

$$g(x_1) - g(x_2) \in L(X)(x_1 - x_2),\quad x_1, x_2 \in X \in \mathbb{I}B.$$

In many papers special interval operators for $F$ are described, and questions about existence and uniqueness of a solution $x^*$ or the question: "under which assumptions do we get $X_\infty = x^*$?" are answered.

(Some basic papers of this subject are: [3], [4], [6], [10], [12], [20], [21], [23], [24], [25], [27], [28] and [29]. See also the references of [13].)

Adams [1] and Gay [8], [9] have extended these studies to the case that g is not exactly known (e. g., if the coefficients of g are intervals). They thereby start from the function $g: B \subseteq \mathbb{R}^n \times D \subseteq \mathbb{R}^p \to \mathbb{R}^n$. If $x^*(d)$ denotes a zero of $g(x,d) = 0$ with a fixed $d \in D$, then they define a set of solutions $X^*$ by $X^* := \{x^*(d) \mid d \in D\}$, and they give bounds or intervals, respectively, which enclose $X^*$.

In another model we use a function strip $G: B \subseteq \mathbb{R}^n \to \mathbb{I}\mathbb{R}^n$ instead of a function g, and instead of a zero we get a zero set $X^*$, which can be enclosed with the help of a fixed interval of an interval operator F, or pseudofixed interval, respectively. (See [7], [14], [15], [16], [17].)

## 2. Notation and basic concepts

Lower case letters denote real values (vectors, matrices and real-valued functions). Capital letters denote sets (interval vectors, interval matrices and interval functions). $\mathbb{I}\mathbb{R}^n$ [or $\mathbb{I}\mathbb{R}^{n\times n}$, respectively] denotes the set of all interval vectors [or interval matrices, respectively], and $\mathbb{I}B := \{X \in \mathbb{I}\mathbb{R}^n \mid X \subseteq B\}$.

If $\Sigma$ is a bounded subset of $\mathbb{R}^n$, we denote by $\square\Sigma := [\inf \Sigma, \sup \Sigma]$ the <u>interval hull</u> of $\Sigma$.

Let $X = [\underline{x},\bar{x}] \in \mathbb{I}\mathbb{R}^n$; then $\operatorname{rad} X := \frac{1}{2}(\bar{x}-\underline{x})$ denotes the <u>radius</u>, $\operatorname{mid} X = \check{x} = \frac{1}{2}(\underline{x}+\bar{x})$ the <u>midpoint</u> and $|X| := \sup(\bar{x},-\underline{x})$ the <u>absolute value</u> of X. Analogous notations apply to $L = [\underline{l},\bar{l}] \in \mathbb{I}\mathbb{R}^{n\times n}$. If $r \in \mathbb{R}^{n\times n}$, then $\sigma(r)$ denotes the spectral radius of r. Concerning interval arithmetic we refer to [5] and [19].

By Neumaier [22] a map $S: \mathbb{I}\mathbb{R}^n \to \mathbb{I}\mathbb{R}^n$ is called <u>sublinear</u> if the following axioms are valid for all $X,Y \in \mathbb{I}\mathbb{R}^n$.

(S1) $X \subseteq Y \Rightarrow SX \subseteq SY$ (inclusion isotonicity),
(S2) $\alpha \in \mathbb{R} \Rightarrow S(X\alpha) = (SX)\alpha$ (homogeneity),
(S3) $S(X+Y) \subseteq SX + SY$ (subadditivity).

We extend S to matrix arguments by applying it to each column of the matrix. Moreover, we set

$$\kappa(S) := Se \quad \text{and} \quad |S| = |SE|,$$

where e denotes the unit matrix, and $E = [-e,e]$. (In [22] the interval matrix $\kappa(S) \in \mathbb{IR}^{n\times n}$ is called the kernel and the nonnegative matrix $|S|$ is called the absolute value of S).

A sublinear map is called normal, if for all $X \in \mathbb{IR}^n$,

(S4) $\quad \mathrm{rad}(SX) \geq |S|\,\mathrm{rad}\,X;$

it is called centered, if

(S5) $\quad X \in \mathbb{IR}^n,\ \mathrm{mid}(SX) = 0 \ \Rightarrow\ \mathrm{mid}\,X = 0,$

and regular, if

(S6) $\quad x \in \mathbb{R}^n,\ 0 \in Sx \ \Rightarrow\ x = 0.$

Let $L \in \mathbb{IR}^{n\times n}$ be a regular interval matrix (i. e., each matrix $l \in L$ is regular). Then $L^{-1}$ is defined by

$$L^{-1} := \square\{l^{-1} \mid l \in L\}.$$

Moreover, a sublinear map $L^I$ is called inverse of L, if

$$l^{-1}x \in L^I X \quad \text{for all} \quad l \in L,\ x \in X.$$

$L := \{L_{ik}\} \in \mathbb{IR}^{n\times n}$ is called H-matrix, if the real matrix $\langle L\rangle := \{l_{ik}\}$ with $l_{ii} := \inf\{|l| \mid l \in L_{ii}\}$ and $l_{ik} := -|L_{ik}|$ for $i \neq k$, $i,k = 1(1)n$, is an M-matrix.

## 3. A function strip and its zero set

Let $G: B \subseteq \mathbb{R}^n \to \mathbb{IR}^n$ be a map which associates with each $x \in B$ an interval

$$G(x) := [\underline{g}(x), \overline{g}(x)]. \tag{1}$$

Such a map is called a function strip. We call

$$X^* := \{x \in B \mid \underline{g}(x) \leq 0 \leq \overline{g}(x)\}$$

the zero set of G (which can be empty).

Remark: This zero set $X^*$ encloses the set of solutions defined by Adams [1] or Gay [9], respectively.

We assume that G on each $X \in \mathbb{I}B$ satisfies an interval Lipschitz condition, i. e., the real functions $\underline{g}$ and $\bar{g}$ both satisfy the same interval Lipschitz condition

$$\left.\begin{array}{l}\underline{g}(x_1) - \underline{g}(x_2) \in L(X)(x_1-x_2), \quad \bar{g}(x_1) - \bar{g}(x_2) \in L(X)(x_1-x_2) \\ \text{for all } x_1, x_2 \in X \in \mathbb{I}B.\end{array}\right\} \tag{2}$$

We call $L: \mathbb{I}B \to \mathbb{I}\mathbb{R}^{n\times n}$ a Lipschitz operator and assume that L is inclusion isotone, i. e.

$$X \subseteq Y \Rightarrow L(X) \subseteq L(Y). \tag{3}$$

## 4. Interval operators of a function strip, properties of such operators and some general theorems

Let the map $F: \mathbb{I}B \to \mathbb{I}\mathbb{R}^n$ be a continuous interval operator. We call $\hat{X} \in \mathbb{I}B$ a fixed interval of F if $F(\hat{X}) = \hat{X}$, and we call $X \in \mathbb{I}B$ a pseudo-fixed interval of F, if $F(X) \supseteq X$.

Properties of an interval operator (see definition in [15]):
Let $X, Y \in \mathbb{I}B$, $X^* \subseteq B$ the zero set of a function strip G and $\hat{X} \in \mathbb{I}B$ a fixed interval of an interval operator F. Then we call F

| | | |
|---|---|---|
| (E1) | inclusion isotone, | if $X \subseteq Y \Rightarrow F(X) \subseteq F(Y)$, |
| (E2) | normal, | if $X^* \subseteq \hat{X}$, |
| (E3) | inclusion preserving, | if $X^* \subseteq X \Rightarrow X^* \subseteq F(X)$, |
| (E4) | fixed interval preserving, | if $\hat{X} \subseteq X \Rightarrow \hat{X} \subseteq F(X)$, |
| (E5) | strong, | if $F(X) \supseteq X \supseteq \hat{X} \Rightarrow X = \hat{X}$. |

The following theorems are valid.

Theorem 1: If the continuous interval operator F is inclusion preserving and $\emptyset \neq X^* \subseteq X_0 \subseteq B$ [or fixed interval preserving and $\hat{X} \subseteq X_0 \subseteq B$, respectively], then the interval sequence $\{X_k\}$ defined by

$$X_{k+1} := X_k \cap F(X_k), \quad k = 0,1,2,\ldots \tag{4}$$

converges; hence

$$\left.\begin{array}{l}\lim_{k\to\infty} X_k = X_\infty \supseteq X^* \quad [\text{or } X_\infty \supseteq \hat{X}, \text{ respectively}] \\ \qquad \text{with} \quad F(X_\infty) \supseteq X_\infty\end{array}\right\} \tag{5}$$

holds.

(See Theorem 2.3 in [15]).

Theorem 2: If the continuous interval operator F is fixed interval preserving and strong, and if F possesses a fixed interval $\hat{X} \subseteq X_0$, then we get for the interval sequence $\{X_k\}$ defined by (4)

$$\lim_{k\to\infty} X_k = \hat{X}.$$

(See Theorem 2.4 in [15]).

Theorem 3: If the continuous interval operator F is inclusion isotone, and if $F(X) \subseteq X$ then there exists a fixed interval $\hat{X}$ of F.

In the following sections we discuss three classes of special interval operators.

## 5. Newton-like interval operators

$$N_0(X) := \check{x} - L^I G(\check{x}), \tag{6}$$

where $L := L(X_0)$ is a constant Lipschitz matrix of (2), and $L^I$ denotes a normal and centered inverse of L.

Theorem 4 (conclusion from Theorem 4.2 in [17]): Let $N_0$ be defined by (6), then $N_0$ is normal and inclusion preserving.

Supposing, $\hat{X} := [\hat{x} - \text{rad}\,\hat{X},\ \hat{x} + \text{rad}\,\hat{X}]$ is a fixed interval of F. Then it follows from (S5) that

$$\text{mid}\, G(\hat{x}) = 0, \tag{7}$$

and from (S4), as well as (3.8) in [17] that

$$\text{rad}\, X = |L^I|\, \text{rad}\, G(\hat{x}). \tag{8}$$

This means: the absolute value (matrix) $|L^I|$ determines the "size" of a fixed interval.

A second "measure" is the matrix

$$\mathrm{rad}(L^I L),$$

which we call convergence matrix, because it is responsible for the speed of convergence of the iteration (4). Moreover, the following statement holds:

Theorem 5: Let $N_0$ be defined by (4). If a fixed interval $\hat{X}$ of $N_0$ exists, and if

$$\sigma(\mathrm{rad}(L^I L)) < 1, \qquad (9)$$

then $N_0$ is a strong operator, i. e., the property (E5) is satisfied. (See Proposition 6.2 in [17]).

Next we discuss four examples of an inverse $L^I$ of L.

1. $L^G Z := \mathrm{IGA}(L,Z)$,

   where IGA denotes the interval Gauss algorithm (see [6]). Sufficient conditions for the existence of $L^G$ are:

   (i) L = regular and n = 2

   (see Reichmann [30]).

   (ii) L = H-matrix

   (see Alefeld [2]).

Generally, the regularity of L is not sufficient for the existence of $L^G$. (See a variant of Reichmann [30] in the remark 3 of Theorem 3 in [22]).

For the following three inverses $L^I$ we assume that the Lipschitz-matrix $L \in \mathbb{IR}^{n\times n}$ is strongly regular, i. e. by (7.1) in [17]:

The matrix

$$a := (\mathrm{mid}\, L)^{-1} \qquad (10)$$

exists, and with

$$r := |a|\,\mathrm{rad}\, L \qquad (11)$$

the condition

$$\sigma(r) < 1 \qquad (12)$$

holds. Then

$$q := (e-r)^{-1}r \qquad (13)$$

exists, and it is a nonnegative matrix.

2. $L^I = L^K$: $L^K Z := aZ + (qE)(aZ)$, (14)

3. $L^I = L^V$: $L^V Z := [e-q, e+q](aZ)$, (15)

4. $L^I = L^P$: $L^P Z := (aL)^G (aZ)$. (16)

Remark: For the one-dimensional case, and if $G$ degenerates to a function g, $L^G$ was introduced by Moore [19] and applied by Nickel [27] and many other authors. Among other things, the multi-dimensional case was discussed by Alefeld/Herzberger [4]. New results for $L^G$ were derived by Neumaier [22] and [26]. For the function strip, $L^K$ was introduced by Krawczyk [14], and $L^V$ by Krawczyk/Neumaier [16]. $L^P$ means preconditioning of L with $(\text{mid}\, L)^{-1}$ (see section 6 in [22]), which was applied by Hansen/Smith [11].

Theorem 6: If $\sigma(q) < 1$, where q is defined by (13), then the inverses $L^K$, $L^V$ and $L^P$ are regular.
(See examples 2, 3 and 4 of section 7 in [17]).

Theorem 7: Let $N_0$ be defined by (6) with $L^I = L^K$ (see (14)). Then $N_0$ is a fixed interval preserving operator.
(See Theorem 5.4 in [15]).

Remark: It is not necessary, however, that $N_0$ with $L^G$ or $L^V$, $L^P$, respectively be fixed interval preserving, as the following example shows:

Let be $\underline{g}(x) := \begin{cases} 4x-6, & \text{if } x \geq 0, \\ 2x-6, & \text{if } x < 0, \end{cases}$ $\quad \overline{g}(x) = 4x+6$.
Then $L = [2, 4]$ and $L^G Z = L^P Z = Z \times \left[\frac{1}{4}, \frac{1}{2}\right]$. $\hat{X} = [-3, 3]$ is a fixed interval of $N_0$ with $L^I = L^G$. Choosing $X_0 = [-3, 5] \supseteq \hat{X}$ we obtain $X_1 = [-3, 2] \not\supseteq \hat{X}$, which is contrary to the statement of Theorem 7.

(As far as $L^V$ is concerned, see example 5.3 in [15]).

Theorem 8: Let $N_0$ be defined by (6) with $L^I = L^K$, and let $\sigma(q) < 1$, where q is defined by (13). Then $N_0$ is a strong operator.
(See Theorem 5.5 in [15]).

Remark: In comparing this result with Theorem 5 we can say: $\sigma(q) < 1$ is a weaker assumption than (9) that is, $\sigma(\mathrm{rad}(L^I L)) < 1$, because $\mathrm{rad}(L^K L) = 2q$.

Comparison of the cases 1., 2. and 3.:

(i) $L^V Z \subseteq L^K Z, \quad L^P Z \subseteq L^K Z$ for all $Z \in \mathbb{IR}^n$, (17)

(ii) $|L^K| = |L^V| = |L^P| = (e-r)^{-1}|a|$, (18)

(iii) $\mathrm{rad}(L^K L) = 2q$, (19)

(iv) $q \leq \mathrm{rad}(L^V L) \leq 2q$, (20)

(v) $q \leq \mathrm{rad}(L^P L) \leq 2q$. (21)

From (17) it follows that the application of $L^V$ and $L^P$ yields better results than the use of $L^K$. However, we cannot tell whether $L^V$ or $L^P$ is more favorable. A comparison with $L^G$ is difficult, because the fixed interval of $N_0^G$ (applying $L^G$ in (6)) generally does not coincide with the fixed interval $N_0^K$ (applying $L^K$ in (6)).

However, it follows from (18) that all interval operators: $N_0^K$, $N_0^V$ and $N_0^P$ possess the same fixed interval $\hat{X}$.

Conclusion from Theorem 8: If the assumptions of Theorem 8 are fulfilled then $N_0$ with $L^I = L^V$ or $L^I = L^P$, respectively, is strong. Because of (17) it follows that $N_0^V(X) \subseteq N_0^K(X)$, as well as $N_0^P(X) \subseteq N_0^K(X)$. ($N_0^V$ denotes the operator (6) with $L^I = L^V$, and $N_0^P$ is the notation if $L^I = L^P$. If a fixed interval $\hat{X}$ of $N_0^K$ exists, then by (18) $\hat{X}$ is a fixed interval of $N_0^V$ and $N_0^P$, too. Therefore $N_0^V(X) \supseteq X \supseteq \hat{X}$ implies $N_0^K(X) \supseteq X \supseteq \hat{X}$, and by applying Theorem 7 we obtain $X = \hat{X}$. Analogously, $N_0^P(X) \supseteq X \supseteq \hat{X}$ implies $X = \hat{X}$.

Remark: Theorem 8 and the conclusions are true only if $L^I$ is constant. However, it is not necessary that the interval operator

$$N(X) := \check{x} - L^I(X)G(\check{x}) \tag{22}$$

with variable $L(X)$ is strong, as the examples 5.4 and 5.5 in [15] show. Furthermore, there can exist more than one fixed interval which all have the same midpoint $\hat{x}$. In contrast, $N_0$ has at most one fixed interval if $\sigma(r) < 1$, since the zero $\hat{x}$ of the equation $\mathrm{mid}\, G(x) = 0$ is unique, and by (8), $\mathrm{rad}\,\hat{X}$ is independent of $X$ (see example 6.1 in [16]). It is even possible that there exists a fixed interval of N but not of

$N_0$ (see example 6.3 in [16]). The contrary statement is not true. If $N_0$ possesses a fixed interval then there exists at least one fixed interval of N (see Theorem 6.5 in [16]). If $\hat{X}_0$ denotes a fixed interval of $N_0$ and $\hat{X}$ a fixed interval of N, then $\hat{X} \subseteq \hat{X}_0$ holds for each fixed interval $\hat{X}$ of N (see Theorem 6.4 in [16]).

Overestimation: Let $X^* \neq \emptyset$, then the iteration method (4) with the operator (6) yields a limit interval $X_\infty \supseteq X^*$ (Theorem 1) or $\hat{X} \supseteq X^*$ (Theorem 2), respectively. The "distance" of the interval hull of X* from $\hat{X}$ can be bounded by the following Theorem.

Theorem 9: Let $L^I$ be a regular and centered inverse of L. Suppose that for each $l \in L$ the inequality

$$|L^I| \leq |l^{-1}| + 2\,\mathrm{rad}\,(\kappa(L^I)) \tag{23}$$

holds. Then it follows that

$$0 \leq \mathrm{rad}\,\hat{X} - \mathrm{rad}\,\square X^* \leq 2(\mathrm{rad}(L^{-1}) + \mathrm{rad}(\kappa\,(L^I)))\,\mathrm{rad}\,G(\hat{x}) \tag{24}$$

(see Theorem (5.1), (iv) in [17]).

Remarks: 1. The assumption (23) is valid for $L^I = L^K, L^V, L^P$.

2. Since $L^{-1} \subseteq \kappa(L^I)$, the bound (24) can be simplified by

$$\mathrm{rad}\,\hat{X} - \mathrm{rad}\,\square X^* \leq 4\,\mathrm{rad}\,(\kappa(L^I))\,\mathrm{rad}\,G(\hat{x}).$$

3. If $\mathrm{rad}(\kappa\,(L^I)) = O(\varepsilon)$ and $\mathrm{rad}G(\hat{x}) = O(\varepsilon)$ then it follows from (24) that $\mathrm{rad}\,\hat{X} - \mathrm{rad}\,\square X^* = O(\varepsilon^2)$. This means quadratic vonvergence, if $\varepsilon \to 0$.

## 6. K-operators

Instead of the Newton-like interval oprator (6) for iteration (4) we can use the operator

$$K_0(X) := \check{x} - aG(\check{x}) + (rE)(X-\check{x}), \tag{25}$$

where a and r are defined by (10) and (11).

If we assume (12) - such that $\sigma(r) < 1$ - then there exists at most one fixed interval $\hat{X}$ of $K_0$. By setting $\hat{X} = [\hat{x} - \mathrm{rad}\,\hat{X},\ \hat{x} + \mathrm{rad}\,\hat{X}]$ we obtain

$$\mathrm{mid}\,G(\hat{x}) = 0, \quad \mathrm{rad}\,\hat{X} = (e-r)^{-1}|a|\,\mathrm{rad}\,G(\hat{x}). \tag{26}$$

From (8) and (18) it follows that a fixed interval of $N_0$ with $L^I = L^K, L^V, L^P$ coincides with a fixed interval of $K_0$. With respect to the properties of $K_0$ the following theorem holds.

Theorem 10: Under the assumption (12), the interval operator $K_0$ defined by (25) is inclusion isotone, normal, inclusion preserving, fixed interval preserving and strong.
(See Theorem 5.1 - 5.5 in [15].)

Remarks: 1. For the statement: "$K_0$ is a strong operator" the assumption $\sigma(q) < 1$ is not necessary. In contrast, $\sigma(q) < 1$ is necessary for $N_0$ to be a strong operator.
(See example 5.4 in [15].)

2. The remark referring to the property: "strong" and to fixed intervals of (22) with $L^I(X) = L^K(X), L^V(X), L^P(X)$ yields an analogous result for the interval operator

$$K(X) := \check{x} - a(X)G(\check{x}) + (r(X)E)(X-\check{x}) \tag{27}$$

with $a(x) := (\text{mid}\, L(X))^{-1}$ and $r(X) := |a(X)|\, \text{rad}\, L(X)$.
Each fixed interval of $N(X)$ is a fixed interval of $K(X)$, too, and vice versa.

In correspondence with the bound (24) with regard to the distance of the solution set $X^*$ from a fixed interval $\hat{X}$, the inequality

$$0 \leqq \text{rad}\,\hat{X} - \text{rad}\,X^* \leqq (2\,\text{rad}\,(L^{-1}) + q|a|)\,\text{rad}\,G(\hat{x}) \tag{28}$$

holds.

By comparing the bound (28) with (24) we obtain from (24) in the case $L^I = L^V$, because of $\kappa(L^I) = [a - q|a|,\, a + q|a|]$ (see example 3, (iv) in [17]),

$$\text{rad}\,\hat{X} - \text{rad}\,X^* \leqq (2\,\text{rad}\,L^{-1} + 2q|a|)\,\text{rad}\;G(\hat{x}),$$

which is less favorable then (28).

## 7. The optimal operator

Under special assumptions we can apply an operator $\mathcal{O}_0$ or $\mathcal{O}$ instead of $N_0$ or N, respectively, $K_0$ or K, which optimally encloses a generalized zero set.

Assumption: Let a matrix $b \in \mathbb{R}^{n \times n}$ exist such that

$$0 \leq e - bL(X) \quad \text{for all} \quad X \in \mathbb{IB} \tag{29}$$

holds.

b can be split as $b = b^+ - b^-$, where $b^+ := \sup(b,0)$, $b^- := \sup(-b,0)$. Let

$$\underline{f}(x) := x - b^+\overline{g}(x) + b^-\underline{g}(x),$$
$$\underline{f}(x) := x - b^+\underline{g}(x) + b^-\overline{g}(x).$$

Then the optimal operator is given by

$$\mathcal{O}_0(X) := [\underline{f}(\underline{x}), \ \overline{f}(\overline{x})]. \tag{30}$$

Remark: If $b$ = constant then $\mathcal{O}_0$ is indenpendent of $L(X)$.

Theorem 11: If the assumption (29) holds, then the operator $\mathcal{O}_0$ defined by (30) is inclusion isotone, normal, inclusion preserving and fixed interval preserving. If, in addition,

$$\sigma|e - bL(X)| < 1 \quad \text{for all} \quad X \in \mathbb{IB}, \tag{31}$$

then $\mathcal{O}_0$ is a strong operator.
(The proof of this theorem will be published later).

We call the set

$$X^{**} := \{x \in D \mid \underline{f}(x) \leq x \leq \overline{f}(x)\}$$

a pseudo-zero set; note that

$$X^* \subseteq X^{**}$$

holds.

Theorem 12: Let the assumptions (29) and (31) be fulfilled. If, additionally, $X^* \neq \emptyset$, and a fixed interval $\hat{X}$ of $\mathcal{O}_0$ exists, then the iterated sequence (4) with the operator (30) converges, and we obtain

$$\lim_{k\to\infty} X_k = \hat{X} = \square X^{**}.$$

Theorem 13 (existence): Under the assumptions (29), (31) and $\mathcal{O}_0(X) \subseteq X$ there exists a fixed interval $\hat{X} = \square X^{**}$ of $\mathcal{O}_0$.

Theorem 14 (existence): Under the assumptions of Theorem 13, if $X^{**} \neq \emptyset$ and $\square X^{**} \subseteq \operatorname{int} B$ *) there exists a fixed interval $\hat{X} = \square X^{**}$ of $\mathcal{O}_0$.

Theorem 15 (overestimation): Let $X^* \neq \emptyset$ and a fixed interval of $\mathcal{O}_0$ exist, $\bar{s} := |e-bL(\hat{X})|$, $t = 2(e-\bar{s})^{-1}$ and $z := \operatorname{rad} G(\hat{x}) + \operatorname{rad} L(\hat{X}) \operatorname{rad} \hat{X}$. Then

$$\operatorname{rad} \hat{X} - \operatorname{rad} \square X^* \leqq \inf\{tb^+z,\ tb^-z\} \qquad (32)$$

holds.

Special cases: $b^- = 0$ or $b^+ = 0$: Then it follows from (32) that

$$\hat{X} = \square X^*,$$

i.e., the zero set $X^*$ can be enclosed optimally by $\hat{X}$.

Remark: Let $L(X)$ be inverse nonnegative for all $X \in \mathbb{IB}$. By choosing $b = \bar{l}^{-1}(X_0) = b^+$, $b^- = 0$ we obtain the operator $\mathcal{O}_0(X) = [\underline{x} - \bar{l}^{-1}\bar{g}(\underline{x}), \bar{x} - \bar{l}^{-1}\underline{g}(\bar{x})]$. If $b(X) = \bar{l}^{-1}(X)$ is variable, we then get the interval operator $\mathcal{O}(X)$ which was introduced in [7] (see (4.1) in [7]). In contrast to $N_0$ and $N$, or $K_0$ and $K$, respectively the fixed interval of $\mathcal{O}_0$ coincides with the fixed interval of $\mathcal{O}$, such that under the given assumptions there exists at most one fixed interval of $\mathcal{O}$.

If the function strip $G$ degenerates to a real function $g$ then we obtain the method of Li [18].

*) $\operatorname{int} B$ means the interior of $B$.

REFERENCES


[1] Adams, E.: On Sets of Solutions of Collections of Nonlinear Systems in $\mathbb{IR}^n$. Interval Mathematics 1980, ed. by K. Nickel. Academic Press, New York-London-Toronto, 247 - 256 (1980).

[2] Alefeld, G.: Über die Durchführbarkeit des Gauss'schen Algorithmus bei Gleichungen mit Intervallen als Koeffizienten. Computing Suppl. 1, 15 - 19 (1977).

[3] Alefeld, G.: On the Convergence of some Interval-Arithmetic Modifications of Newton's Method. SIAM J. Num. Anal. 21, 363 - 372 (1984).

[4] Alefeld, G. and Herzberger, J.: Über das Newton-Verfahren bei nichtlinearen Gleichungssystemen. ZAMM 50, 773 - 774 (1970).

[5] Alefeld, G. and Herzberger, J.: Introduction to Interval Computations. Academic Press, New York, 1983.

[6] Alefeld, G. and Platzöder, L.: A Quadratically Convergent Krawczyk-like Algorithm. SIAM J. Num. Anal. 20, 210 - 219 (1983).

[7] Garloff, J. and Krawczyk, R.: Optimal Inclusion of a Solution Set. Freiburger Intervall-Berichte 84/8, 13 - 33 (1984).

[8] Gay, D. M.: Pertubation Bounds for Nonlinear Equations. SIAM J. Num. Anal. 18, 654 - 663 (1981).

[9] Gay, D. M.: Computing Pertubation Bounds for Nonlinear Algebraic Equations. SIAM J. Num. Anal. 20, 638 - 651 (1983).

[10] Hansen, E.: Interval Forms of Newtons Method. Comp. 20, 153 - 163 (1978).

[11] Hansen, E. and Smith, R.: Interval Arithmetic in Matrix Computations, Part II. SIAM J. Num. Anal. 4, 1 - 9 (1967).

[12] Krawczyk, R.: Newton-Algorithmen zur Bestimmung von Nullstellen mit Fehlerschranken. Comp. 4, 187 - 201 (1969).

[13] Krawczyk, R.: Intervalliterationsverfahren. Freiburger Intervall-Berichte 83/6 (1983).

[14] Krawczyk, R.: Interval Iteration for Including a Set of Solutions. Comp. 32, 13 - 31 (1984).

[15] Krawczyk, R.: Properties of Interval Operators. Freiburger Intervall-Berichte 85/3, 1 - 20 (1985).

[16] Krawczyk, R. and Neumaier, A.: An Improved Interval Newton Operator. Freiburger Intervall-Berichte 84/4, 1 - 26 (1984).

[17] Krawczyk, R. and Neumaier, A.: Interval Newton Operators for Function Strips. Freiburger Intervall-Berichte 85/7, 1-34 (1985).

[18] Li, Q. Y.: Order Interval Newton Methods for Nonlinear Systems. Freiburger Intervall-Berichte 83/8 (1983).

[19] Moore, R. E.: Interval Analysis. Prentice-Hall, Inc. Englewood Cliffs, N. J. 1966.

[20] Moore, R. E.: A Test for Existence of Solutions to Nonlinear Systems. SIAM J. Numer. Anal. 14, 611 - 615 (1977).

[21] Moore, R. E.: New Results on Nonlinear Systems. Interval Mathematics 1980, ed. by K. Nickel, 165 - 180 (1980).

[22] Neumaier, A.: New Techniques for the Analysis of Linear Interval Equations. Linear Algebra Appl. 58, 273 - 325 (1984).

[23] Neumaier, A.: An Interval Version of the Secant Method. BIT 24, 366 - 372 (1984).

[24] Neumaier, A.: Interval Iteration for Zeros of Systems of Equations. BIT 25, 256 - 273 (1985).

[25] Neumaier, A.: Existence of Solutions of Piecewise Differentiable Systems of Equations. Freiburger Intervall-Berichte 85/4, 27 - 34 (1985).

[26] Neumaier, A.: Further Results on Linear Interval Equations. Freiburger Intervall-Berichte 85/4, 37 - 72 (1985).

[27] Nickel, K.: On the Newton Method in Interval Analysis. MRC Technical Summary Report # 1136, University of Wisconsin, Madison (1971).

[28] Nickel, K.: A Globally Convergent Ball Newton Method. Comp. 24, 97 - 105 (1980).

[29] Qi, L. Q.: A Generalization of the Krawczyk-Moore Algorithm. 'Interval Mathematics 1980', ed. by K. Nickel. Academic Press, 481 - 488 (1980).

[30] Reichmann, K.: Abbruch beim Intervall - Gauss-Algorithmus. Comp. 22, 355 - 361 (1979).

# Arbitrary Accuracy with Variable Precision Arithmetic

Fritz Krückeberg

Gesellschaft für Mathematik und Datenverarbeitung mbh (GMD)

Summary

For the calculation of interval-solutions $Y$ including the true solution $y$ of a given problem we need not only that $y \epsilon Y$ holds. Furthermore we are interested in the value of $span(Y)$. So we should get that for an a priori and arbitrarily given bound $\varepsilon > 0$ the calculation yields that the error remains below $\varepsilon$ or that $span(Y) < \varepsilon$. It is possible to realize $span(Y) < \varepsilon$ for arbitrary $\varepsilon > 0$ by using an interval–arithmetic with variable word length within a three–layered methodology, including validation/verification of the solution. The three–layered methodology consists of

- Computer algebra procedures,
- the numerical algorithm,
- an interval arithmetic with variable and controllable word length.

Examples are given in the domain of linear equations and ordinary differential equations (initial value problems).

## 1. General Aspects

For the calculation of interval–solutions $Y$ including the true solution $y$ of a given problem we need not only that

$$y \,\epsilon\, Y$$

holds. Furthermore we are interested in the value of $span(Y)$. So we should get that for an a priori and arbitrarily given bound $\varepsilon > 0$ the calculation yields that the error remains below $\varepsilon$ or that

$$span(Y) < \varepsilon. \tag{1}$$

It is possible to realize (1) for arbitrary $\varepsilon > 0$ by using a combination of some methods that are described in this paper.
If we realize (1), a next level of Interval Mathematics is reached: in addition to the inclusion relation $y \epsilon Y$ the 'quality' $\varepsilon$ can be fulfilled in a way, that the result is better (but not too much better) than $\varepsilon$. This means that the amount of computation can be restricted in dependency on the quality $\varepsilon$ which is wanted. But how such flexibility may be reached?

## 2. Interval–arithemetic with variable word length

The condition (1) needs flexibility in two directions: to be accurate enough if necessary and to be rough enough if possible. By introducing an interval–arithmetic with variable word length this flexibility can be arrived.

For this reason we have defined a special interval–arithmetic with variable word length on a decimal base $b = 10^4$ with a fixed point architecture. Principally it would be possible to use a floating point arithmetic with variable mantissa length. The restriction to a fixed point arithmetic with variable word length is not a principal restriction.
The used interval–arithmetic was realized by FORTRAN subroutines. These subroutines leads to the four basic operations $+, -, *, /$. The result of each operation leads to a new fixed point expression, but with a longer word length if needed (to save all digits of the result). If the word length is growing more than needed an interval rounding can be executed to an arbitrary number of digits (depending on the precision needed).

For these operations some additional parameters are stored at the end of the fixed point word. The parameters contain the number of leading digits on the left side of the decimal point and the number of digits needed after the decimal point. Figure 1 explains the architecture of such a fixed point word:

| | | | | | | | | | | | | |
|---|---|---|---|---|---|---|---|---|---|---|---|---|
| | ... | $c_2$ | $c_1$ | $c_0$ | . | $c_{-1}$ | $c_{-2}$ | $c_{-3}$ | $c_{-4}$ | ... | $m$ | $\pm n$ |

*Figure* 1

$c_i$ are digits with $-(10^4 - 1) \leq c \leq +(10^4 - 1)$,
each digit must be interpreted as an integer (with a sign $\pm$).

$m =$ the number of significant digits after the decimal point,

$\pm n =$ the number of significant digits on the left side of the decimal point
(if $n < 0$, $n$ denotes the number of zeros after the decimal point).

The minimal value $z$ of such a fixed point word is then defined by

$$z = \sum_{i=-m}^{n-1} c_i \, 10^{4i}$$

To restrict the amount of computation the values of $m$ and $n$ are limited by

$$\begin{array}{rcccl} 0 & \leq & m & \leq & 20 \\ -19 & \leq & n & \leq & 20 \end{array}$$

but the limits are not fixed, they can be varied without changing the subroutines.

The interval–arithmetic with variable word length uses a pair of such words (with variable parameters $m$ and $n$). In consistency with this variable base subroutines for a set of elementary functions are written (sine, cosine, natural logarithm, exponential function). For example it is possible to calculate

$$Y := \sin Z \qquad (Y, Z \text{ are intervals}) \tag{2}$$

in such a way that

$$span(Y) < \varepsilon$$

for an arbitrarily given $\varepsilon > 0$ if $span(Z)$ is small enough to realize (2). The amount of computation time to calculate such 'variable' subroutines as sine depends on the value of $\varepsilon$. If $\varepsilon$ is smaller the calculation time is longer. It is not easy to write such subroutines but they are needed to get the flexibility that is wanted.

## 3. The 3 Levels of computations

By introducing an interval–arithmetic with variable word length only the lowest level I of a concept of (see Figure 2) computation is realized. We have to control the variable word length in a dynamic way during the computation time by the numerical methods that are used. So the numerical methods should contain control parameters for the control of word length. A control parameter of the given method may be for example the step size $h$ of a numerical integration method. If the step size $h$ is smaller the word length should be longer. The numerical methods are placed at the level II of our 3–level diagram. Computer–algebra–systems belongs to the level III. The interconnection of all three levels is described in the following diagram (Figure 2):

Now it is possible to realize (1) for given problems belonging to some problem classes: Computer-algebra-systems are used for formal computation to avoid rounding errors as long as possible or to find out well conditioned domains for the application of numerical methods at level II. Between the numerical methods and the basic operations at level I exist a control feedback to determine the word length (dynamically) or to determine the number of iterations (or other parameters) at level II. In a similiar way there will be a control feedback between level II and level I (the number of higher order derivatives of a given function at level I may be connected with the number of iterations that are needed at level II).
By using such a 3-level-control structure several examples for different problem classes are calculated by the author and some cooperating persons [Demmel, Leisen, Pascoletti].

| THE 3 LEVELS OF WORK | PROCEDURES | PROCESS-CONTROL | |
|---|---|---|---|
| LEVEL OF SYMBOLIC COMPUTATION | COMPUTER-ALGEBRA-SYSTEMS AND REDUMA | 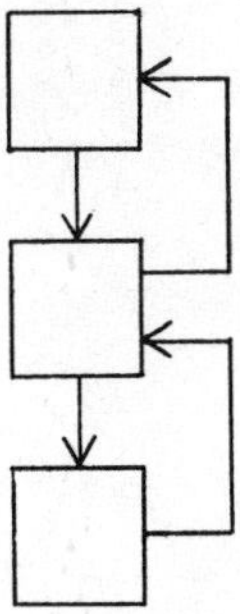 | |
| LEVEL OF NUMERICAL METHODS | NUMERICAL METHODS | | VERIFICATION OF RESULTS |
| LEVEL OF ARITHMETIC OPERATIONS | DYNAMIC INTERVAL ARITHMETIC ARITHMETIC WITH VARIABLE WORD LENGTH | | |

*Figure 2*

## 4. Examples

### 4.1 Initial value problems for ordinary differential equations

The initial value problem

$$y' = f(x,y), \qquad x\,\epsilon[\underline{a},\overline{a}]\,\epsilon II(I\!R), \quad y\epsilon I\!R^n$$

$$y(\underline{a}) = s, \qquad s\epsilon I\!R^n$$

is to be solved for all values $x$ with

$$\underline{a} \leq x \leq \overline{a}$$

and the true solution

$$y(x)$$

should be included in a stepwise interval polynomial $Y(x)$ so that

$$y(x)\epsilon Y(x) \qquad for\,all\,x\,with \quad \underline{a} \leq x \leq \overline{a} \tag{3}$$

and the condition

$$span(Y(x)) < \varepsilon \qquad (4)$$

holds for an arbitrarily a priori given $\varepsilon > 0$.
We now give a general description of our method [ see Leisen, Krückeberg ]. To construct a solution for this problem the functions $f(x,y)$ are to be described in the form of a formal expression similar to expressions that are used in Computer–algebra. Then it is possible to calculate the derivatives of the functions $f(x,y)$ in a formal way by recursive methods that are typical for Computer–algebra routines.

The formal calculation of derivatives is a typical process within level III of Figure 2. The stepwise integration starts with

$$x_0 = \underline{a}$$

and leads to a sequence of points

$$x_0,\ x_1,\ x_2,\ \ldots$$

with a variable step size $h_i$. If it is possible to include the solution $y(x_{i-1})$ in an interval vector$[u_{i-1}]$ then we have to realize the integration from $x_{i-1}$ to $x_i$ so that the calculated interval vektor $[u_i]$ includes $y(x_i)$. The integration step from $x_{i-1}$ to $x_i$ is shown in Figure 3.

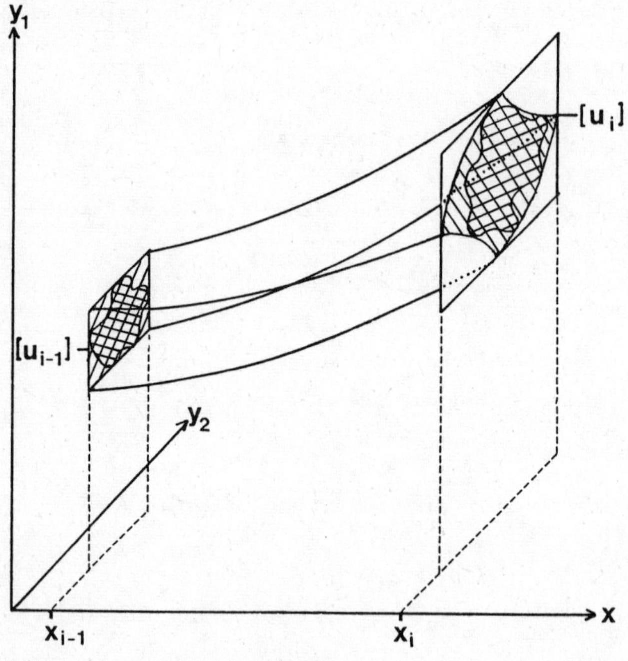


*Figure* 3

To held condition (4) it is necessary to control the step size $h$. But $h$ is only one parameter to contol the precision. Annother parameter is the degree $s$ of the highest derivative of $f(x,y)$ that is used in the taylor evaluation for the integration steps. But the parameters $h$ and $s$ are not sufficient to fulfill condition (4). We also need a variable word length (that means we have to control $m$). If we use a control procedure for all three parameters

$h$ step size

$s$ degree of derivation

$m$ word length of interval–arithmetic

then it is possible to fulfill condition (4). Many examples are calculated [Krückeberg, Leisen] with good results.

Examples

For a linear system of differential equations

$$\begin{array}{rcrcrcrl} y_1' & = & \frac{3}{2}y_1 & - & y_2 & - & \frac{1}{2}y_3, & \quad y_1(0) = 2 \\ y_2' & = & -\frac{1}{2}y_1 & + & 2y_2 & + & \frac{1}{2}y_3, & \quad y_2(0) = -6 \\ y_3' & = & \frac{1}{2}y_1 & + & y_2 & + & \frac{5}{2}y_3, & \quad y_3(0) = 0 \end{array}$$

and two different conditions (4) with

$$\varepsilon_1 = 10^{-50}$$

and

$$\varepsilon_2 = 10^{-70}$$

the integration was executed. Figure 4 shows the results. For the $\varepsilon - axis$ a logarithmical scale is used.

$$|\tilde{y} - y|$$

denotes the deviation from the true solution.

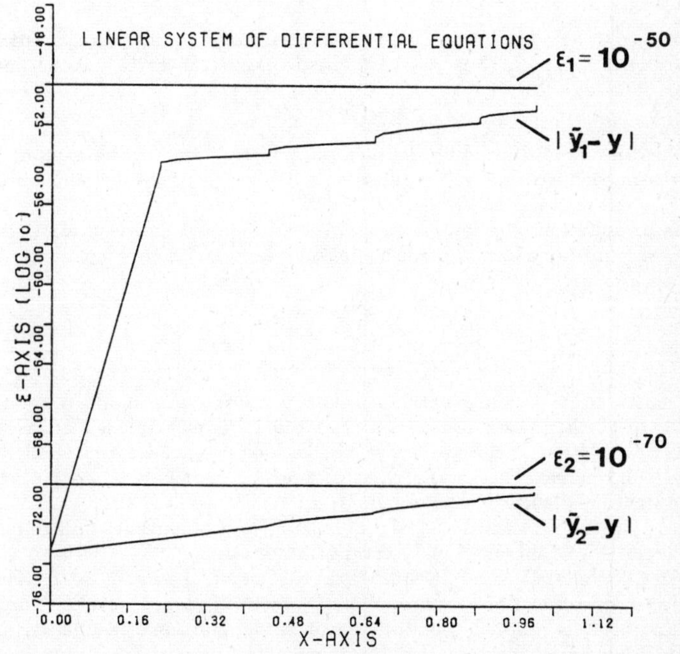


*Figure 4*

Figure 4 shows the possibility to control the integration in such a way that condition (4) is fulfilled. The values of $\varepsilon_1, \varepsilon_2$ for practical use are extremely small, but it should be explained that variable word length arithmetic is powerful enough to fulfill such conditions. The method was also tested for several nonlinear systems of differential equations. Further work is to be done for other problems in the domain of differential equations, specially for boundary value problems.

### 4.2 Systems of linear equations

If a system of linear equations

$$Ax = b \tag{5}$$

is given then we ask for an interval vector $X$ so that

$$x \epsilon X$$

and

$$span(X) < \varepsilon \tag{6}$$

is fulfilled for an arbitrarily given $\varepsilon > 0$. To solve this problem we have constructed a procedure that includes a control process for the word length. By [Demmel, Krückeberg] such a procedure was constructed and tested.

In some cases only integer solutions of (5) are of interest. Then it is sufficient to take

$$\varepsilon = 0.5$$

and to control the word length in such a way that the result is not much better than $\varepsilon$. The constructed algorithm uses only as much precision as needed to achieve (6).

#### Examples

To test our algorithm we tried to invert Hilbert matrices scaled to have integer entries. For Hilbert matrices of the order 10, 11, 12, 13 it was easy to calculate an interval vector solution $X$ with

$$span(X) < 10^{-9}$$

in such a way that the algorithm adjust its own parameters to minimize computation time. It is obviously that the needed word length (controlled by the algorithm) depends on the condition of the matrix.
In the case of linear equation only level II and level I of Figure 2 are used. But sometimes it may be of interest to perform some formal transformations (level III) before the numerical algorithm is started.

## 5. Consequences

It seems to be realistic to follow the general aspects of chapter 1 and to fulfill condition (1). Then a next level of Interval Mathematics is reached and an economic principle is introduced: to minimize computation time in relation to the given $\varepsilon > 0$. The including of Computer–algebra (level III of Figure 2) seems to be necessary in many cases of analytic problems to have a better chance for controlling the numerical algorithm at level II.
Hopefully several new methods will be constructed to solve problems of applied mathematics (and of Interval Mathematics) at this level of Interval Mathematics.
The new methods will contain several parameters at the level I, II and III. So it may be problematic to realize an automatic control of all parameters by feedback loops between the levels I, II and III, so that condition (1) is fulfilled. For this reason some additional feedback is needed between the user and the computer: expert system are expected to be useful to support such a feedback between the user and the computation process.

## REFERENCES

Demmel J W, Krückeberg F (1985) An Interval Algorithm for Solving of Linear Equations to Prespecified Accuracy. Computing 34: 117–129.

Krier N, Spelluci P (1975) Untersuchungen der Grenzgenauigkeit von Algorithmen zur Auflösung linearer Gleichungssysteme mit Fehlererfassung. In: Interval Mathematics, Lecture Notes in Computer Science 29: 288–297 ed. by Nickel K, Springer Verlag, Berlin Heidelberg, New York.

Krückeberg F, Leisen R (1985) Solving Initial Value Problems of Ordinary Differential Equations to Arbitrary Accuracy with Variable Precision Arithmetic. In: Proceedings of the 11th IMACS World Congress on System Simulation and Scientific Computation: 111–114 ed. by Wahlstrom B, Henriksen R, Sundby N P, Oslo, Norway.

Leisen R (1985) Zur Erzielung variabel vorgebbarer Fehlereinschließungen für gewöhnliche Differentialgleichungen mit Anfangswertmengen mittels dynamisch steuerbarer Arithmetik. Diplomarbeit, Universität Bonn.

Pascoletti K H (1982) Ein intervallanalytisches Iterationsverfahren zur Lösung von linearen Gleichungssystemen mit mehrparametriger Verfahrenssteuerung. Diplomarbeit, Universität Bonn.

Prof. Dr. Fritz Krückeberg
Gesellschaft für Mathematik und Datenverarbeitung mbh (GMD)
Postfach 1240, Schloß Birlinghoven, D–5205 St. Augustin

# AN INTERVAL METHOD FOR SYSTEMS OF ODE

S.Markov and R.Angelov
Bulgarian Academy of Sciences, Sofia, and
High Institute for Economics, Varna, Bulgaria

Abstract. Considered is an interval algorithm producing bounds for the solution of the initial value problem for systems of ordinary differential equations $\dot{x}(t)=f(t,c,x(t))$, $x(t_0)=x_0$, involving inexact data c, $x_0$, taking values in given intervals $C=[\underline{c}, \overline{c}]$, resp. $X_0=[\underline{x}_0, \overline{x}_0]$. An estimate for the width of the computed inclusion of the solution set is given under the assumption that f is Lipschitzian. In addition, if f is quasi-isotone, the computed bounds converge to the interval hull of the solution set and the order of global convergence is O(h).

1.Notations. As usually, we denote by I(R) the set of all compact intervals on the real line R of the form $A=[\underline{a}, \overline{a}]$ ; $V_n(I(R))$ means the set of all n-dimentional interval vectors on R of the form $([\underline{a}_1,\overline{a}_1], \ldots, [\underline{a}_n, \overline{a}_n])$. The width of $A=[\underline{a}, \overline{a}]$, $B=[\underline{b}, \overline{b}] \in I(R)$ is denoted by $w(A) = \overline{a} - \underline{a}$, the joint of A and B is denoted by $A \vee B = [\min\{\underline{a}, \underline{b}\}, \max\{\overline{a}, \overline{b}\}]$.

2.Formulation of the problem. We consider the initial value problem for systems of n ODE's:

$$\text{(1a)} \qquad \dot{x} = f(t,c,x(t)), \quad x(t_0)=x_0$$

involving inexact (interval) data for the parameter vector c and the initial condition vector $x_0$, that is

$$\text{(1b)} \qquad c \in C = [\underline{c}, \overline{c}] \in V_m(I(R)), \quad x_0 \in X_0 = [\underline{x}_0, \overline{x}_0] \in V_n(I(R)).$$

We shall seek an enclosure $[\underline{s}, \overline{s}]$ of the set $\{\check{x}\}$ of all solutions of (1) on an interval $T=[t_0,\overline{t}]$ (assuming that all solutions $\check{x}$ of (1) does exist on T), that is $\underline{s}(t) \le \check{x}(t) \le \overline{s}(t)$ for every solution $\check{x}$ of (1) and every $t \in T$.

We shall assume that f is an n-vector function defined on $T \times C \times D$, $D=([\underline{d}_1,\overline{d}_1], \ldots, [\underline{d}_n,\overline{d}_n])$, such that f is continuous with respect to c, Lipschitzian with respect to t and x, and quasiisotone with respect to x.

The algorithm described below requires the effective computation of:
a) the intervals $f_i(t,C,x)=\{f_i(t,c,x):c\in C\}$, $i=1,\dots,n$, for every $t\in T$, $x\in D$; the end-points of these intervals will be further denoted by $\underline{f}_i(t,x)$, resp. $\overline{f}_i(t,x)$;
b) the intervals $\underline{F}_i(\tilde{T},\tilde{D}):=\{\underline{f}_i(t,x):t\in\tilde{T},x\in\tilde{D}\}$, $i=1,\dots,n$, and the intervals $\overline{F}_i(\tilde{T},\tilde{D}):=\{\overline{f}_i(t,x):t\in\tilde{T},x\in\tilde{D}\}$, $i=1,\dots,n$, for every $\tilde{T}\subset T,\tilde{D}\subset D$.

3. Description of the algorithm. Let $h>0$ be a sufficiently small step, defining a mesh $t_k=t_o+kh\in T$, $k=0,1,\dots,\hat{k}$. The bounds $\underline{s}(t)=(\underline{s}_1(t),\dots,\underline{s}_n(t))$, $\overline{s}(t)=(\overline{s}_1(t),\dots,\overline{s}_n(t))$ for the solution set $\{\check{x}\}$ are seeked in the form of polygones with vertices at the mesh points $t_k$.
We set $\underline{s}(t_o)=\underline{x}_o$, $\overline{s}(t_o)=\overline{x}_o$. Assuming that $\underline{s}$, $\overline{s}$ are already computed at some $t_k$, that is $\underline{s}(t_k)$, $\overline{s}(t_k)$ are such that $\underline{s}_i(t_k)\le\check{x}_i(t_k)\le\overline{s}_i(t_k)$, $i=1,2,\dots,n$, we then compute $\underline{s}$, $\overline{s}$ in the interval $T_k=[t_k,\ t_{k+1}]$ by means of the following iteration procedure:
i) for the upper bound $\overline{s}$ we have for $r=0,1,2,\dots,\mathfrak{r}$

$$\overline{Z}_i^{(0)}=[\underline{d}_i,\ \overline{d}_i],\quad i=1,\dots,n,$$
$$[\overline{p}_i^{(r)},\ \overline{q}_i^{(r)}]=\overline{F}_i(T_k,\overline{Z}_1^{(r)},\overline{Z}_2^{(r)},\dots,\overline{Z}_n^{(r)}),\quad i=1,\dots,n,$$
$$\overline{Z}_i^{(r+1)}=\ \overline{s}_i(t_k)\vee(\overline{s}_i(t_k)+\overline{p}_i^{(r)}h)\vee(\overline{s}_i(t_k)+\overline{q}_i^{(r)}h)\quad,\ i=1,\dots,n,$$
$$\overline{s}_i(t)=\overline{s}_i(t_k)+\overline{q}_i^{(\mathfrak{r})}(t-t_k),\quad t\in T_k,\ i=1,\dots,n;$$

ii) for the lower bound $\underline{s}$ we compute for $r=0,1,\dots,\mathfrak{r}$

$$\underline{Z}_i^{(0)}=[\underline{d}_i,\ \overline{d}_i],\quad i=1,\dots,n,$$
$$[\underline{p}_i^{(r)},\ \underline{q}_i^{(r)}]=\underline{F}_i(T_k,\underline{Z}_1^{(r)},\underline{Z}_2^{(r)},\dots,\underline{Z}_n^{(r)}),\quad i=1,\dots,n,$$
$$\underline{Z}_i^{(r+1)}=\ \underline{s}_i(t_k)\vee(\underline{s}_i(t_k)+\underline{p}_i^{(r)}h)\vee(\underline{s}_i(t_k)+\underline{q}_i^{(r)}h)\quad,\ i=1,\dots,n,$$
$$\underline{s}_i(t)=\underline{s}_i(t_k)+\underline{p}_i^{(\mathfrak{r})}(t-t_k),\quad t\in T_k,\ i=1,\dots,n.$$

Theorem. For any nonnegative integer $r=0,1,2,\dots$

$$\underline{Z}_i^{(r+1)}\subset\underline{Z}_i^{(r)},\qquad\overline{Z}_i^{(r+1)}\subset\overline{Z}_i^{(r)}.$$

Proof. We have for $r=0$

$$\underline{Z}_i^{(1)}=\ \underline{s}_i(t_k)\vee(\underline{s}_i(t_k)+\underline{p}^{(0)}h)\vee(\underline{s}_i(t_k)+\underline{q}_i^{(0)}h)\quad,\ i=1,\dots,n.$$

Since $\underline{s}_i(t_k)\in(\underline{d}_i,\ \overline{d}_i)$, we can take $h$ sufficiently small so that $\underline{Z}_i^{(1)}\subset[\underline{d}_i,\ \overline{d}_i]=\underline{Z}_i^{(0)}$. Assume that $\underline{Z}_i^{(r)}\subset\underline{Z}_i^{(r-1)}$ for some $r\ge 2$. Then, since $\underline{F}_i$ is inclusion isotone, we have

$$[\underline{p}_i^{(r)},\ \underline{q}_i^{(r)}]=\ \underline{F}_i\ (T_k,\ \underline{Z}_1^{(r)},\dots,\underline{Z}_n^{(r)})$$
$$\subset\ \underline{F}_i(T_k,\underline{Z}_1^{(r-1)},\dots,\underline{Z}_n^{(r-1)})=[\ \underline{p}_i^{(r-1)},\underline{q}_i^{(r-1)}].$$

Thus, $\underline{p}_i^{(r-1)} \leq \underline{p}_i^{(r)} \leq \underline{q}_i^{(r)} \leq \underline{q}_i^{(r-1)}$ and therefore

$$\underline{s}_i(t_k)+\underline{p}_i^{(r-1)}h \leq \underline{s}_i(t_k)+\underline{p}_i^{(r)}h,$$

$$\underline{s}_i(t_k)+\underline{q}_i^{(r-1)}h \geq \underline{s}_i(t_k)+\underline{q}_i^{(r)}h,$$

that is, $\underline{Z}_i^{(r+1)} \subset \underline{Z}_i^{(r)}$.

The inclusion $\overline{Z}_i^{(r+1)} \subset \overline{Z}_i^{(r)}$ is proved analogously.

We shall now prove that $\underline{s}$, $\overline{s}$ are bounds for the solution set.

Theorem. $\underline{s}(t) \leq \{\check{x}(t)\} \leq \overline{s}(t), \quad t \in T_k$.

Proof. For any nonnegative integer r (and, in particular r=$\mathfrak{r}$) we have

$$\underline{s}_i'(t)=\underline{p}_i^{(r)} \leq \underline{f}_i(t,x_1,\ldots,x_n),\ t \in T_k,\ x_j \in \underline{Z}_j^{(r)},\ j=1,\ldots,n$$

Since $\underline{Z}_j^{(r+1)} \subset \underline{Z}_j^{(r)}$, we have for every $t \in T_k$

$$\underline{s}_j'(t)=\underline{s}_j(t_k)+\underline{p}_j^{(r)}(t-t_k) \in \underline{Z}_j^{(r+1)} \subset \underline{Z}_j^{(r)},\ j=1,\ldots,n.$$

Therefore, $\underline{s}_i'(t) \leq \underline{f}_i(t,\underline{s}_1(t),\ldots,\underline{s}_n(t)), \quad i=1,\ldots,n.$

Analogously, it can be shown that

$$\overline{s}_i'(t) \geq \overline{f}_i(t,\overline{s}_1(t),\ldots,\overline{s}_n(t)), \quad i=1,\ldots,n.$$

Let $\check{x}(t)$ be be an arbitrary solution of (1) corresponding to some $c \in C$ and $x_0 \in X_0$. We have

$$\underline{s}'(t) \leq \underline{f}(t,\underline{s}(t)) \leq f(t,c,\underline{s}(t)),$$

$$\overline{s}'(t) \geq \overline{f}(t,\overline{s}(t)) \geq f(t,c,\overline{s}(t)),$$

$$\underline{s}(t_0)=\underline{x}_0 \leq x_0 \leq \overline{x}_0=\overline{s}(t).$$

From the relations

$$\underline{s}'(t) \leq f(t,c,\underline{s}(t)), \qquad \check{x}'(t)=f(t,c,\check{x}(t)), \qquad \overline{s}'(t) \geq f(t,c,\overline{s}(t)),$$

$$\underline{s}(t_0) \leq x_0, \qquad \check{x}(t_0)=x_0, \qquad \overline{s}(t_0) \geq x_0,$$

assuming that f is quasiisotone in x, we obtain $\underline{s}(t) \leq \check{x}(t) \leq \overline{s}(t)$, according to an well known theorem of M.Müller [2].

Remark. The inclusions $[\overline{p}^{(r+1)}, \overline{q}^{(r+1)}] \subset [\overline{p}^{(r)}, \overline{q}^{(r)}]$ , $[\underline{p}^{(r+1)}, \underline{q}^{(r+1)}] \subset [\underline{p}^{(r)}, \underline{q}^{(r)}]$ show that the the computed bounds of the solution set are improved at each step of the iteration procedure. Each step produces a local (that is in the interval $T_k$) approximation of the solution set of order $O(h^2)$; thus $\mathfrak{r}$=2 is a suitable choice for practical applications. If computer arithmetic with directed roundings is available, then the effect of finite convergence can be recommended as stopping criteria of the local iteration procedure [1].

4. An estimate for the width of the obtained inclusion. For any $\tilde{T} \subset T$, $X_i \subset [\underline{d}_i, \overline{d}_i]$, i=1,...,n, we have

$$w(\underline{F}_j(\tilde{T},X_1,\dots,X_n)) = \max_{\substack{t \in \tilde{T},\\ x_i \in X_i, i=1,\dots,n}} \underline{f}_j(t,x_1,\dots,x_n) - \min_{\substack{t \in \tilde{T},\\ x_i \in X_i, i=1,\dots,n}} \underline{f}_j(t,x_1,\dots,x_n)$$

$$= \underline{f}_j(t',x_1',\dots,x_n') - \underline{f}_j(t'',x_1'',\dots,x_n'') \leq$$

(assuming that f is Lipschitzian in t and x)

$$\leq \tilde{L}\,|t'-t''| + \sum_{i=1}^{n} L_i\,|x_i'-x_i''|$$

$$\leq \tilde{L}\,w(\tilde{T}) + \sum_{i=1}^{n} L_i w(X_i)$$

where $\tilde{L}, L_1,\dots,L_n$ are some constants. Analogously we have

$$w(\overline{F}_j(\tilde{T},X_1,\dots,X_n)) \leq \tilde{L}\,w(\tilde{T}) + \sum_{i=1}^{n} L_i w(X_i)$$

for any $\tilde{T} \subset T$, $X_i \subset [\underline{d}_i, \overline{d}_i]$, i=1,...,n.

The above estimates are used in the proof of the following

Theorem. The bounds $\underline{s}$, $\overline{s}$ for the solution set $\{\check{x}\}$ of problem (1) satisfy the inequality

$$\overline{s}_i(t)-\underline{s}_i(t) \leq A_1 w_0 + A_2 M + A_3 h, \quad i=1,\dots,n, \quad t \in T,$$

wherein

$$w_0 = w(X_0) = \max_{i=1,\dots,n} |\overline{x}_{0i} - \underline{x}_{0i}|, \quad M = \max_{\substack{t\in T, x\in D,\\ i=1,\dots,n}} (\overline{f}_i(t,x) - \underline{f}_i(t,x)),$$

and the constants $A_1$, $A_2$, $A_3$ do not depend on $w_0$, M and h.

Proof. Let $\overline{s}_i(t_k)-\underline{s}_i(t_k) \leq w_k$, i=1,...,n, k=0,1,...,$\tilde{k}$. We have

$$\overline{s}_i(t_{k+1})-\underline{s}_i(t_{k+1}) = \overline{s}_i(t_k)+\overline{q}_i^{(\xi)}h-\underline{s}_i(t_k)-\underline{p}_i^{(\xi)}h$$

$$= \overline{s}_i(t_k)-\underline{s}_i(t_k) + (\overline{q}_i^{(\xi)}-\underline{p}_i^{(\xi)})h$$

$$\leq w_k + hw(\overline{F}_i(T_k,\overline{Z}_1^{(\xi)},\dots,\overline{Z}_n^{(\xi)}) \vee \underline{F}_i(T_k,\underline{Z}_1^{(\xi)},\dots,\underline{Z}_n^{(\xi)}))$$

$$\leq w_k + h(w(\overline{F}_i(T_k,\overline{Z}_1^{(\xi)},\dots,\overline{Z}_n^{(\xi)})) + w(\underline{F}_i(T_k,\underline{Z}_1^{(\xi)},\dots,\underline{Z}_n^{(\xi)}))$$

$$+h\,|\overline{f}_i(t_k,\overline{s}(t_k)) - \underline{f}_i(t_k,\underline{s}(t_k))|$$

$$\leq w_k+h(\tilde{L}(t_{k+1}-t_k)+\sum_{j=1}^{n}L_j w(\overline{Z}_j^{(\xi)})+\tilde{L}(t_{k+1}-t_k)+\sum_{j=1}^{n}L_j w(\underline{Z}_j^{(\xi)})$$

$$+h(\,|\overline{f}_i(t_k,\overline{s}(t_k))-\overline{f}_i(t_k,\underline{s}(t_k)|\;+\;|\overline{f}_i(t_k,\underline{s}(t_k))-\underline{f}_i(t_k,\underline{s}(t_k))|\,)$$

$$\leq w_k+h(2\tilde{L}h+\sum_{j=1}^{n}L_j(w(\overline{Z}_j^{(\xi)})+w(\underline{Z}_j^{(\xi)})))$$

$$+hw(\overline{F}_i(t_k,[\,\overline{s}(t_k)\vee\underline{s}(t_k)]))+hM$$

$$\leq w_k+2\tilde{L}h^2+h\sum_{j=1}^{n}L_j(w(\overline{Z}_j^{(\xi)})+w(\underline{Z}_j^{(\xi)}))+h\sum_{j=1}^{n}L_j w_k+Mh$$

$$=(1+hL)w_k+2\tilde{L}h^2+h\sum_{j=1}^{n}L_j(w(\overline{Z}_j^{(\xi)})+w(\underline{Z}_j^{(\xi)}))+Mh,\text{ where } L=\sum_{i=1}^{n}L_i.$$

We have $w(\underline{Z}_j^{(\xi)})\leq(\underline{s}_j(t_k)+|\underline{q}_j^{(\xi-1)}|h)-(\underline{s}_j(t_k)-|\underline{p}_j^{(\xi-1)}|h)$

$$=(|\underline{q}_j^{(\xi-1)}|+|\underline{p}_j^{(\xi-1)}|)h\leq 2Gh,\text{ wherein } G=\max_{\substack{t\in T,x\in D,\\ c\in C,i=1,\dots,n}}|f_i(t,c,x)|.$$

Similarly, we obtain $w(\overline{Z}_j^{(\xi)})\leq 2Gh$.

We thus have $\overline{s}_i(t_{k+1})-\underline{s}_i(t_{k+1})\leq w_{k+1}=(1+hL)w_k+Mh+(2\tilde{L}+4GL)h^2, i=1,\dots,n.$

From the equalities

$$w_1=(1+hL)w_0+Mh+(2\tilde{L}+4GL)h^2,$$
$$w_2=(1+hL)w_1+Mh+(2\tilde{L}+4GL)h^2,$$
$$\dots\dots\dots\dots\dots$$
$$w_k=(1+hL)w_{k-1}+Mh+(2\tilde{L}+4GL)h^2$$

we obtain

$$w_k=(1+hL)^k w_0+hM\sum_{j=0}^{k-1}(1+hL)^j+h^2(2\tilde{L}+4GL)\sum_{j=0}^{k-1}(1+hL)^j$$

We have $h\leq(\overline{t}-t_0)/\tilde{k}\leq(\overline{t}-t_0)/k$, $k=1,\dots,\tilde{k}$, and, therefore

$$(1+hL)^k\leq(1+(\overline{t}-t_0)L/k)^k\leq e^{(\overline{t}-t_0)L}=A_1,$$

$$h\sum_{j=0}^{k-1}(1+hL)^j=((1+hL)^k-1)/L\leq(e^{(\overline{t}-t_0)L}-1)/L=A_2.$$

Finally, we obtain for $w_k$

$$w_k\leq A_1w_0+A_2M+A_3h,\quad A_3=A_2(2\tilde{L}+4GL),\quad k=1,2,\dots,\tilde{k}.$$

From the relation $\overline{s}_i(t_k)-\underline{s}_i(t_k)\leq w_k\leq A_1w_0+A_2M+A_3h$, $i=i,\dots,n$, for $k=1,2,\dots,\tilde{k}$ and the fact that $\overline{s}_i(t)$ and $\underline{s}_i(t)$ are polygones on T we may conclude that $\overline{s}_i(t)-\underline{s}(t) \leq A_1 w_0+A_2M+A_3h$, $i=1,\ldots,n, t \in T$, which proves the theorem. Let us remark that in the above estimate $w_0$ and M can be considered as measures for inexactness of the initial condition data, resp. of the right-hand side f of problem (1).

5.Convergence. Convergence of the enclosure $[\underline{s},\overline{s}]$ to the interval hull of the solution set $\text{hull}\{\check{x}\} = [\inf\{\check{x}\}, \sup\{\check{x}\}]$ can be demonstrated under the assumption that there exist $c_1$, $c_2 \in C$, such that

$$\underline{f}(t,x)=f(t,c_1,x), \qquad \overline{f}(t,x)=f(t,c_2,x), \qquad t \in T,\ x \in D.$$

Denote by $\underline{x}$, $\overline{x}$ the solutions of the initialvalue problems(2), resp.(3)

$$(2) \qquad \dot{x} = \underline{f}(t,x), \quad x(t_o) = \underline{x}_o,$$

$$(3) \qquad \dot{x} = \overline{f}(t,x), \quad x(t_o) = \overline{x}_o$$

Let $\check{x}$ be an arbitrary solution of (1) corresponding to some $c \in C$, $x_o \in X_o$. We then have $\underline{x}' = \underline{f}(t,\underline{x}) \leq f(t,c,\underline{x})$, $\overline{x}' = \overline{f}(t,\overline{x}) \geq f(t,c,\overline{x})$, $\underline{x}_o \leq x_o \leq \overline{x}_o$, and, by the quasi-isotonicity of f, $\underline{x}(t) \leq \check{x}(t) \leq \overline{x}(t)$. Since $\underline{x}(t)$, $\overline{x}(t)$ are solutions (belong to $\{\check{x}\}$ ), we have

$$\text{hull}\{\check{x}\} = [\underline{x}(t), \overline{x}(t)].$$

Apply the algorithm to problems (2) and (3) and denote by $\underline{u}_h, \overline{u}_h$, and $\underline{v}_h$, $\overline{v}_h$ the corresponding bounds for $\underline{x}$, resp. $\overline{x}$, produced by the algorithm. Since $\underline{x}$, $\overline{x}$ are solutions of particular exact problems, we have $\overline{u}_h-\underline{u}_h \to 0$, $\overline{v}_h-\underline{v}_h \to 0$ with O(h). The functions $\underline{u}_h$ and $\overline{v}_h$ are bounds for $\text{hull}\{\check{x}\}$. The relations $\underline{x} - \underline{u}_h \leq \overline{u}_h - \underline{u}_h \to 0$, $\overline{v}_h - \overline{x} \leq \overline{v}_h - \underline{v}_h \to 0$ show that the computed bounds $\underline{u}_h$, $\overline{v}_h$ tend to $\text{hull}\{\check{x}\}$ with O(h).

A FORTRAN program realizing the above algorithm is available (as part of a program package called RINA).

## REFERENCES

1.R.E.Moore.Methods and Applications of Interval Analysis. SIAM Studies in Applied Mathematics. 1979.

2.W.Walter.Differential and Integral Ineqalities. Springer, 1970.

# LINEAR INTERVAL EQUATIONS

A. Neumaier
Institut für Angewandte Mathematik
Universität Freiburg
Freiburg i. Br.
West Germany

Abstract. This is a short survey of theory and techniques for the solution of linear interval equations with square or rectangular coefficient matrix.

Notation. $\mathbb{IR}^n$ denotes the set of interval vectors with n components, and $\mathbb{IR}^{m\times n}$ the set of interval m×n-matrices; an interval matrix containing only one element is called thin. In this survey we shall use the terms vector and matrix as synonyms for interval vector and interval matrix. The midpoint, radius, and absolute value of a matrix $A \in \mathbb{IR}^{m\times n}$ are understood componentwise and denoted by $\check{A} = \text{mid}\,A$, $\rho(A) = \text{rad}\,A$, and $|A|$, respectively. Similar definitions apply for vectors. In the following, $A \in \mathbb{IR}^{m\times n}$ is a fixed interval matrix, and $b \in \mathbb{IR}^n$ a fixed interval vector.

A linear interval equation with coefficient matrix A and righthand side b is defined as the family of linear equations

$$\tilde{A}\tilde{x} = \tilde{b} \quad (\tilde{A} \in A,\ \tilde{b} \in b); \tag{1}$$

the solution set of (1) is the set

$$\Sigma(A,b) := \{\tilde{x} \in \mathbb{R}^n \mid \tilde{A}\tilde{x} = \tilde{b} \text{ for some } \tilde{A} \in A,\ \tilde{b} \in b\}.$$

By a result of Beeck [6], the solution set can also be described as

$$\Sigma(A,b) = \{\tilde{x} \in \mathbb{R}^n \mid A\tilde{x} \cap b \neq \emptyset\}. \tag{2}$$

The criterion $\tilde{x} \in \Sigma(A,b)$ iff $A\tilde{x} \cap b \neq \emptyset$ is equivalent to a famous perturbation theorem of Oettli and Prager [26].

The solution set $\Sigma(A,b)$ is bounded if A is <u>regular</u>, i.e. if all matrices $\tilde{A} \in A$ have rank n. A sufficient condition for the regularity of a square matrix A is (Ris [29]):

$$\check{A}^{-1} \text{ exists and } |\check{A}^{-1}|\rho(A) \text{ has spectral radius } < 1; \tag{3}$$

let us call such matrices <u>strongly regular</u>. Since the solution set of a linear interval equation may be very complicated, we are interested in finding interval enclosures for $\Sigma(A,b)$. The interval vector with smallest radius containing $\Sigma(A,b)$ is the <u>hull</u> of the solution set,

$$A^H b := \square\Sigma(A,b) = [\inf \Sigma(A,b), \sup \Sigma(A,b)]. \tag{4}$$

## 1. The square case

In this section, we treat the case of a square coefficient matrix (i.e. $m = n$).

<u>1.1 Computing the hull</u>. The computation of $A^H b$ is, in general, a very difficult problem; the known algorithms seem to have a worst case complexity exponential in n. However, for $n = 1$, the hull is computable by simple division, $A^H b = b/A$, and for $n = 2$, a simple method is described in Apostolatos and Kulisch [4]. For general n, several algorithms have been given by Rohn [30], [31], [32]; the algorithm of the first paper is iterative and assumes that A is strongly regular, the other two papers treat the case of general regular A. The algorithms are very time-consuming for large n; moreover, in their present form, they do not account for rounding errors in the computation.

For dimensions n larger than about 5, practical methods are available only in special cases. If A is thin then $A^H b = A^{-1}b$ (Beeck [7]), and if A is an M-matrix then Gauss-Seidel iteration yields the hull (Barth and Nuding [5]). In case that the righthand side b satisfies one of the conditions $b \geq 0$, $b \leq 0$, or $b \ni 0$, the hull can be computed for inverse positive A as $A^H b = [\overline{A}^{-1}, \underline{A}^{-1}]b$ (Beeck [7]); and for the more special case of M-matrices, Gauss elimination gives the hull for these righthand sides (Barth and Nuding [5], Beeck [7]). Further methods for the computation of the hull for M-matrices and inverse positive matrices are described in Beeck [8] and Neumaier [24].

For other matrices we are, at present, restricted to the use of methods which do not compute an optimal enclosure.

1.2 Gauss elimination. We denote by $A^G b$ the result of Gauss elimination applied without pivoting to the linear interval equation (1). Depending on the type of the coefficient matrix, the results may be very good or very bad.

Gauss elimination is almost optimal if A is an M-matrix: We have $A^G b \subseteq [\overline{A}^{-1}, \underline{A}^{-1}]b$ (Neumaier [24]), and in special cases we get the hull (Barth and Nuding [5],Beeck [7] for $b \geqq 0$, $b \leqq 0$, or $b \ni 0$, and Schwandt [37] for thin A). But we warn that standard column pivoting may destroy the M-matrix property, leading to a loss of upto 3 decimals in accuracy (Schätzle [34])

Gauss elimination without pivoting is also reliable for diagonally dominant matrices (experiments of Kopp [17]) and Hessenberg matrices with a special sign structure (Reichmann [27]); it also works well with pivoting for $n = 2$ (Alefeld [3] and Reichmann [28] show that $A^G b$ exists for $n = 2$ iff A is regular).

Gauss elimination can also be performed without pivoting if A is an H-matrix (Alefeld [2]), and with pivoting if A is regular and $\rho(A)$ is sufficiently small (Neumaier [24]). However rounding errors and dependency may lead in these cases to catastrophic overestimation (exponential in n) or even to breakdown due to division by an interval containing zero (Wongwises [40], Schätzle [34]).

Gauss elimination can be coupled with iteration by splitting A as $A = A_0 - E$ and considering the iteration

$$x^{\ell+1} := A_0^G (Ex^{\ell} + b); \qquad (5)$$

cf. Alefeld [3] for $A_0 = \check{A}$, Mayer [21] for more general splittings. A fixpoint $x^\infty$ of (5) is an enclosure for $A^H b$. If this iteration converges for all $x^O$,b then, by Neumaier [24], Gauss elimination cannot break down, and, if $\check{b} = O$, rad $A^G b \leq$ rad $x^\infty$ . However, if A is an M-matrix and $A_O = \bar{A}$ then the iteration converges for all $x^O$,b, and the limit is the hull, $x^\infty = A^H b$.

1.3 Gauss-Seidel iteration. If an initial enclosure $x^O$ for $A^H b$ is known, a nested sequence of enclosures $x^\ell$ for $A^H b$ can be defined by Gauss-Seidel iteration with componentwise intersection,

$$x^{\ell+1} := \Gamma(A,b,x^\ell) \quad (\ell = 0,1,2,\ldots), \tag{6}$$

where the vector $y := \Gamma(A,b,x)$ is defined by

$$\left.\begin{array}{l} y_i' := (b_i - \sum_{k<i} A_{ik}y_k - \sum_{k>i} A_{ik}x_k)/A_{ii}, \\ y_i := x_i \cap y_i' \end{array}\right\} \quad (i = 1,\ldots,n);$$

cf. Ris [29], Neumaier [23]. Clearly the method applies only when $0 \notin A_{ii}$ $(i = 1,\ldots,n)$, although it can be modified for the general case (cf. Hansen and Sengupta [14]).

If A is an M-matrix then the iteration (6) converges to the hull $A^H b$ (Barth and Nuding [5]); if A is an H-matrix it can at least be shown that the limit $x^\infty$ is contained in a vector $A^F b$ independent of the initial enclosure $x^O$ (Neumaier [23]), thus guaranteeing that at least very bad enclosures will be improved. Gay [12] proved that Gauss-Seidel iteration is faster and has a smaller limit radius than the whole-step iteration

$$x^{\ell+1} = x^\ell \cap (b + (I - A)x^\ell);$$

more generally, Neumaier [23] proved the same optimality result within the class of iterations defined by triangular splittings. In particular, overrelaxation cannot improve the iteration (cf. Mayer [22], Cornelius [9]). Recently, based on ideas of Alefeld [1], a symmetric Gauss-Seidel iteration was discussed by Schwandt [35] and Shearer and Wolfe [38] in a nonlinear setting; this iteration is still faster than (6).

An easy way to get an initial enclosure for arbitrary H-matrices A uses Ostrowski's comparison matrix <A> (see Neumaier [23]) and the implication

$$u > 0,\ \langle A\rangle u \geqq v > 0 \Rightarrow A^H b \subseteq [-u,u]\ \|b\|_v\ , \tag{7}$$

where $\|b\|_v = \max\{\,|b_i|/v_i \mid i = 1,\dots,n\}$. If $\beta := \|I - A\|_\infty < 1$ then one can take $v = e := (1,\dots,1)^T$, $u = (1-\beta)^{-1}e$; in general, since $u^* := \langle A\rangle^{-1}e$ satisfies $u^* > 0$ and $\langle A\rangle u^* = e > 0$, any sufficiently good approximation $u$ of $u^*$ leads with $v = \langle A\rangle u$ to a valid bound.

<u>1.4 Preconditioning</u>. To improve the performance of Gauss elimination, Hansen and Smith [15] suggest to precondition (1) by multiplying with a matrix C (they use an approximate inverse of $\check{A}$), leading to the preconditioned system

$$\tilde{A}\tilde{x} = \tilde{b} \quad (\tilde{A} \in CA,\ \tilde{b} \in Cb). \tag{8}$$

The preconditioning with $C = \check{A}^{-1}$ leads to a regular system (8) iff A is strongly regular (Ris [29]); in this case, CA is an H-matrix, and Gauss elimination can be performed with the matrix CA. More generally, if CA is an H-matrix for some matrix C then A must be strongly regular, and if

$$\beta := \|I - CA\| < 1 \tag{9}$$

in some scaled maximum norm then $\beta$ takes its minimal value for the choice $C = \check{A}^{-1}$ (Neumaier [23]). Unless $\beta > 1$ or $\beta$ is very close to 1, the overestimation inherent in the transformation from (1) to (8) is small since by Neumaier [25],

$$\|\mathrm{rad}\ (CA)^H(Cb)\| \leqq \frac{1+\beta}{1-\beta}\|\mathrm{rad}\ A^H b\|\ . \tag{10}$$

If A is inverse positive and $C = \overline{A}^{-1}$ then we even have $(CA)^H(Cb) = A^H b$ (Neumaier [24]).

For $\beta \ll 1$, CA is almost the identity, and it is faster to compute an enclosure for the solution set of (8) by iteration. The oldest method (cf. Krawczyk [18]) uses

$$x^{\ell+1} := x^\ell \cap (Cb + (I - CA)x^\ell), \tag{11}$$

where the initial enclosure $x^0$ is found as

$$x^0 := [-u,u] \quad (u_i = \frac{1}{1-\beta} \|Cb\|_\infty, \; i = 1,\dots,n),$$

assuming that (9) holds in the maximum norm. Every $x^\ell$ is an enclosure of $A^H b$, and by Neumaier [25], who improved a similar result of Gay [12], the radius of the limit $x^\infty$ is still bounded by the righthand side of (10). Ris [29] observed that it is better to use Gauss-Seidel iteration in place of (11); we mentioned already that Gauss-Seidel iteration is indeed faster, and leads to a limit with smaller radius. For the special choice $C = \check{A}^{-1}$, Proposition 2.5 of Krawczyk and Neumaier [19] implies that $(CA)^G(Cb)$ has a still smaller radius; but the improvement is slight if $\beta \ll 1$.

To get least significant bit accuracy for thin A and b, Rump [33] proceeds slightly differently. He computes a sufficiently accurate approximation $\tilde{x}$ of $x^* = A^{-1}b$, constructs the smallest machine-representable interval x containing $\tilde{x}$, and uses the implication

$$Cb + (I - CA)x \subseteq \operatorname{int} x \Rightarrow A \text{ regular}, \; A^H b \subseteq x, \tag{12}$$

a consequence of Brouwer's fixpoint theorem, to check whether x really contains the solution $x^*$.

<u>1.5 Options for sparse M-matrices</u>. For sparse matrices, the inverse is generally full (Duff et al. [11]), and preconditioning is too time and/or space consuming. Also, Gaus-Seidel iteration is much too slow for most practical sparse problems. At present, no fast and reliable method is known for sparse linear interval equations whose coefficient matrix is neither an M-matrix nor diagonally dominant; the apparently quite general methods proposed by Rump [33] and Hahn et. al. [13] suffer from the overestimation problem of Gauss elimination (except for thin problems where the overestimation can be counteracted by using sufficiently accurate multiprecision approximations of the solution). We survey here the M-matrix case; all methods (but not the optimality results) generalize to H-matrices, and in particular to the diagonally dominant case. The methods known to me are:

a) Gauss elimination for matrices with small bandwidth or profile. This is almost optimal for sparse M-matrices, and the standard profile optimization algorithms can be used.

b) Gauss elimination coupled with the iteration (5); see Mayer [21], Schwandt [37]. If $A_O$ is sparse and thin and $A = A_O - E$ is an M-matrix with $A \leqq A_O$, then the iteration converges to the hull (Neumaier [24]).

c) Iteration with incomplete factorizations; see Mayer [20]. This iteration is closely related to method b) but makes more flexible use of the zero pattern.

d) An interval Buneman algorithm; see Schwandt [36]. This algorithm applies to a restricted class of interval equations related to certain elliptic partial differential equations.

e) Aposteriori enclosure of a good approximate solution by Brouwer's fixpoint theorem (Rump [33]) or perturbation theorems (Hahn et al. [13]). The methods proposed are efficient only for special classes of coefficient matrices like M-matrices or diagonally dominant matrices; they behave badly e.g. on thin banded triangular H-matrices A with $A_{ii} = 3$, $A_{ii-1} = 4$, $A_{ii-2} = 5$ and $A_{ik} = 0$ for $k \neq i,i-1,i-2$, especially when the righthand side is not thin.

f) Modified Gauss-Seidel iteration. The iteration starts with the initial enclosure $x^O := [-u,u]\,\|b\|_v$ from (7) and uses an accompanying approximate iteration $\tilde{x}^\ell$ ($\ell = 1,2,\ldots$) converging to $\check{A}^{-1}\check{b}$. A sequence of enclosures $x^\ell$ for $A^H b$ is found as

$$\left.\begin{array}{ll} x^{\ell+1/2} & := \Gamma(A,b,x^\ell) \\ x^{\ell+1} & := x^{\ell+1/2} \cap (\tilde{x}^{\ell+1} + [-u,u]\,\|b - A\tilde{x}^{\ell+1}\|_v) \end{array}\right\} (\ell=0,1,2,\ldots). \quad (13)$$

This iteration converges for M-matrices A to the hull $A^H b$ independently of the choice of the "forcing sequence" $\tilde{x}^\ell$; however, the convergence speed is at least that of the forcing sequence. Thus, any fast-converging iteration scheme for the approximate solution of sparse linear equations can be used to speed up the process. It is also possible to use only one or two steps of (13) aposteriori, i.e. with a good approximation $\tilde{x}^1$ to $\check{A}^{-1}b$.

## 2. The rectangular case

In this section we treat the case of a rectangular coefficient matrix. We restrict ourselves to the overdetermined case (i.e. $m > n$); for the underdetermined case cf. Rump [33]. The problem was posed first by Jahn [16] who suggested Gauss elimination to solve the equation; however, by the same reasons as for the square case this is reliable only for $n = 2$.

2.1 The least squares hull. All other authors concerned with interval methods for overdetermined systems replace the problem of finding $A^H b$ by that of finding the least squares hull

$$A^L b := \square\{\tilde{x} \in \mathbb{IR}^n \mid \tilde{A}^T\tilde{A}\tilde{x} = \tilde{A}^T\tilde{b} \text{ for some } \tilde{A} \in A,\ \tilde{b} \in b\}, \tag{14}$$

and reduce this problem to the square case by observing that

$$A^H b \subseteq A^L b \subseteq x \text{ , where } \binom{r}{x} = \begin{pmatrix} I & A \\ A^T & 0 \end{pmatrix}^H \binom{b}{0} \tag{15}$$

(Spellucci and Krier [39], Rump [33]) or

$$A^H b \subseteq A^L b \subseteq (A^T A)^H (A^T b) \tag{16}$$

(Deif [10]). Unless one has specific reasons to work with $A^L b$ in place of $A^H b$, the approach via (15), (16) is not recommended since $A^L b$ depends very sensitively on scaling and since $\begin{pmatrix} I & A \\ A^T & 0 \end{pmatrix}$ and $A^T A$ are generally much more ill-conditioned than $A$ (squared condition number).

In the thin case, where usually $A^H b = \emptyset$ and the original problem (1) makes no sense, $A^L b$ can often be computed in spite of ill-conditioning by using a sufficiently accurate approximate least squares solution $\tilde{x}$ and a residual form of (15); conditioning problems are reduced by the use of a highly precise arithmetic. See Rump [33].

2.2 Preconditioning. With an approximation $C$ of a pseudo inverse of $\check{A}$ one gets

$$A^H b \subseteq (CA)^H (Cb). \tag{17}$$

If the rank of $\check{A}$ is n and $\rho(A)$ is sufficiently small then CA is a square n×n-matrix, almost the identity. Therefore, the methods discussed in Section 1.4 can be applied to enclose the righthand side of (17). C is computed in a stable way as $C = R^{-1}Q^TD^{-1}$ from an orthogonal decomposition of the diagonally scaled matrix $D\check{A} = QR$; here $D \in \mathbb{R}^{m\times n}$ is the diagonal scaling matrix, $R \in \mathbb{R}^{n\times n}$ is upper triangular, and $Q \in \mathbb{R}^{m\times n}$ consists of the first n columns of a square orthogonal m×m-matrix (Q,Q').

Since the righthand side of (17) is defined even if the system (1) is inconsistant, a consistency check is useful. By (2), a sufficient condtion for consistency is $A\tilde{x} \cap b \neq \emptyset$ for an approximate solution $\tilde{x}$; on the other hand,

$$A^Hb \subseteq x,\ (a^TA)x \cap a^Tb = \emptyset \Rightarrow A^Hb = \emptyset \qquad (18)$$

holds for all $a \in \mathbb{R}^m$. If inconsistency of (18) is suspected one should check (18) for several choices of a; suitable vectors are e.g. the columns of $D^{-1}Q'$.

References.


1. G. Alefeld, Das symmetrische Einzelschrittverfahren bei linearen Gleichungen mit Intervallen als Koeffizienten, Computing 18, 329-340 (1977).

2. G. Alefeld, Über die Durchführbarkeit des Gaußschen Algorithmus bei Gleichungen mit Intervallen als Koeffizienten, Computing Suppl. 1, 15-19 (1977).

3. G. Alefeld, Intervallanalytische Methoden bei nichtlinearen Gleichungen, Jahrbuch Überblicke Mathematik 1979, (ed. S.D. Chatterji et al.), Bibl. Inst.,Mannheim-Wien-Zürich, 63-78 (1979).

4. N. Apostolatos und U. Kulisch, Grundzüge einer Intervallrechnung für Matrizen und einige Anwendungen, Elektron. Rechenanlagen 10, 73-83 (1968).

5. W. Barth und E. Nuding, Optimale Lösung von Intervallgleichungssystemen, Computing 12, 117-125 (1974).

6. H. Beeck, Über Struktur und Abschätzungen der Lösungsmenge von linearen Gleichungssystemen mit Intervallkoeffizienten, Computing 10, 231-244 (1972).

7. H. Beeck, Zur scharfen Außenabschätzung der Lösungsmenge bei linearen Intervallgleichungssystemen, Z. Angew. Math. Mech. 54, T208-T209 (1974).

8. H. Beeck, Zur Problematik der Hüllenbestimmung von Intervallgleichungssystemen, 'Interval Mathematics', (ed. K. Nickel), Lecture Notes in Computer Science 29, Springer Verlag, 150-159 (1975).

9. H. Cornelius, Untersuchungen zu einem intervallarithmetischen Iterationsverfahren mit Anwendungen auf eine Klasse nichtlinearer Gleichungssysteme, Dissertation, Techn. Univ. Berlin (1981).

10. A. S. Deif, to be published.

11. I. S. Duff, A. M. Erisman, C. W. Gear, and J. K. Reid, Some remarks on inverses of sparse matrices, Techn. Memorandum 51, Math. Comp. Sci. Div., Argonne Nat. Lab., Argonne, Illinois (1985).

12. D. M. Gay, Solving interval linear equations, SIAM J. Numer. Anal. 19, 858-870 (1982).

13. W. Hahn, K. Mohr, and U. Schauer, Some techniques for solving linear equation systems with guarantee, Computing 34, 375-379 (1985).

14. E. Hansen and S. Sengupta, Bounding solutions of systems of equations using interval analysis, BIT 21, 203-211 (1981).

15. E. Hansen and R. Smith, Interval arithmetic in matrix computations, Part II, SIAM J. Numer. Anal. 4, 1-9 (1967).

16. K.-U. Jahn, Eine Theorie der Gleichungssysteme mit Intervallkoeffizienten, Z. Angew. Math. Mech. 54, 405-412 (1974).

17. G. Kopp, Die numerische Behandlung von reellen linearen Gleichungssystemen mit Fehlererfassung für M-Matrizen sowie für diagonaldominante und invers-isotone Matrizen, Diplomarbeit, Inst. f. Prakt. Math. Univ. Karlsruhe (1976).

18. R. Krawczyk Newton-Algorithmen zur Bestimmung von Nullstellen mit Fehlerschranken, Computing 4, 187-201 (1969).

19. R. Krawczyk and A. Neumaier, Interval Newton operators for function strips, Freiburger Intervall-Berichte 85(7), 1-34 (1985).

20. G. Mayer, Enclosing the solution set of linear systems with inaccurate data by iterative methods based on incomplete LU-decompositions, Computing 35, 189-206 (1985).

21. G. Mayer, Comparison theorems for an iterative method based on strong splittings, to appear.

22. O. Mayer, Über intervallmäßige Iterationsverfahren bei linearen Gleichungssystemen und allgemeineren Intervallgleichungssystemen, Z. Angew. Math. Mech. 51, 117-124 (1971).

23. A. Neumaier, New techniques for the analysis of linear interval equations, Linear Algebra Appl. 58, 273-325 (1984).

24. A. Neumaier, Further results on linear interval equations, Freiburger Intervall-Berichte 85(4), 37-72 (1985).

25. A. Neumaier, Overestimation in linear interval equations, Freiburger Intervall-Berichte 85(4), 75-91 (1985).

26. W. Oettli and W. Prager, Compatibility of approximate solution of linear equations with given error bounds for coefficients and right-hand sides, Numer. Math. 6, 405-409 (1964).

27. K. Reichmann, Ein hinreichendes Kriterium für die Durchführbarkeit des Intervall-Gauß-Algorithmus bei Intervall-Hessenbergmatrizen ohne Pivotsuche, Z. Angew. Math. Mech. 59, 373-379 (1979).

28. K. Reichmann, Abbruch beim Intervall-Gauß-Algorithmus, Computing 22, 355-361 (1979).

29. F. N. Ris, Interval analysis and applications to linear algebra, D. Phil. Thesis, Oxford (1972).

30. J. Rohn, An algorithm for solving interval linear systems and inverting interval matrices, Freiburger Intervall-Berichte 82(5), 23-36 (1982).

31. J. Rohn, Solving interval linear systems; Proofs to 'Solving interval linear systems'; Interval linear systems, Freiburger Intervall-Berichte 84(7), 1-14, 17-30, 33-58 (1984).

32. J. Rohn, Some results on interval linear systems, Freiburger Intervall-Berichte 85(4), 93-116 (1985).

33. S. M. Rump, Solving algebraic problems with high accuracy, 'A new approach to scientific computation', (ed. U. W. Kulisch and W. L. Miranker), Academic Press, New York, 51-120 (1983).

34. F. Schätzle, Überschätzung beim Gauß-Algorithmus für lineare Intervallgleichungssysteme, Freiburger Intervall-Berichte 84(3), (1984).

35. H. Schwandt, Schnelle fast global konvergente Verfahren für die Fünf-Punkt-Diskretisierung der Poissongleichung mit Dirichletschen Randbedingungen auf Rechteckgebieten, Dissertation, Techn. Univ. Berlin (1981).

36. H. Schwandt, An interval arithmetic approach for the construction of an almost globally convergent method for the solution of the nonlinear Poisson equation on the unit square, SIAM J. Sci. Statist. Comput. 5, 427-452 (1984).

37. H. Schwandt, Krawczyk-like algorithmus for the solution of systems of nonlinear equations, SIAM J. Numer. Anal. 22, 792-810 (1985).

38. J. M. Shearer and M. A. Wolfe, Some algorithms for the solution of a class of nonlinear algebraic equations, Computing 35,63-72 (1985).

39. P. Spellucci und N. Krier, Untersuchungen der Grenzgenauigkeit von Algorithmen zur Auflösung linearer Gleichungssysteme mit Fehlererfassung, 'Interval Mathematics', (ed. K. Nickel), Lecture Notes in Computer Science 29, Springer Verlag, 288-297 (1975).

40. P. Wongwises, Experimentelle Untersuchungen zur numerischen Auflösung von linearen Gleichungssystemen mit Fehlererfassung, 'Interval Mathematics', (ed. K. Nickel), Lecture Notes in Computer Science 29, Springer Verlag, 316-325 (1975).

# HOW TO FIGHT THE WRAPPING EFFECT

Karl Nickel

Institut für Angewandte Mathematik
Universität Freiburg

Freiburg i. Br.
West Germany

Abstract: The main purpose of this paper is
not to give Theorems, Algorithms, ...,
but to give insight in the cause and the consequences of the wrapping effect and to derive herefrom indications of how to eliminate it.

Notations: Small letters denote real values, vectors and functions of these. Capital letters denote both real matrices and matrix functions and sets of values, vectors and corresponding functions; in particular intervals of such quantities.

## 1. The problem

Considered in what follows is the initial value problem for systems of ordinary differential equations

(1) $u'(t) = f(t,u(t))$ for $t \in I := [0,b]$,

(2) $u(0) = a$,

where $0 < b \in \mathbb{R}$, $a \in \mathbb{R}^n$, $u: I \to \mathbb{R}^n$, $f: I \times \mathbb{R}^n \to \mathbb{R}^n$. It is always assumed that at last one solution $\hat{u}$ exists in I, where $\hat{u}$ is from an appropriate function space. For simplicity only "exact" solutions are treated; i.e. no approximations, no round off errors, no truncation errors are considered. In order to express the dependance of the solution $\hat{u}$ with respect to the initial vector a the notation

$\hat{u}(t;a)$ for the solutions of (1), (2)

is used.
Wanted are set functions $U(t) \subseteq \mathbb{R}^n$ such that

(3) $\hat{u}(t;a) \in U(t)$ for $t \in I$

for all solutions $\hat{u}$ of (1), (2). Herein the set U may be an n-dimensional interval, a ball or another suitable set which is easy to determine.

Problem I:

Let $A \subseteq \mathbb{R}^n$ be a bounded set and replace the initial value condition (2) by the initial inclusion condition

(2') $u(0) = a \in A.$

Wanted are again set functions U(t) such that the inclusion (3) holds for all $a \in A$.

Definition: The inclusion (3) is called optimal under the initial inclusion (2') if

$$\forall\, t \in I\ \forall\, y \in U(t)\ \exists\, a \in A\colon\ y = u(t;a);$$

i.e. if the solutions u(t;a) "fill out completely" the set function U(t) for $a \in A$.

In the Sections 3, 5 and 6 of this paper also the extended

Problem II:

is considered: Let F(t,y) be a bounded set in $\mathbb{R}^n$ for $t \in I$ and $y \in \mathbb{R}^n$ and replace the differential equation (1) by the differential inclusion

(1') $u'(t) \in F(t,u(t))$ for $t \in I.$

To be solved is again the inclusion (3) for all solutions of (1'), (2') and special attention is given again to the above defined optimality.

In the paper [6] the author gave a survey on interval methods for the numerical solution of the problem (1), (2). It contains a list of 123 publications from this field. In it also the wrapping effect is regarded. In the meantime a new publication on this effect appeared by Gambill and Skeel [2].

## 2. Moore's Example

In Figure 1 the well known example of R.E. Moore to the Problem I is sketched. It should be self-explaining and shows that by using intervals U(t) (at the points $t = \pi/6$, $t = \pi/3$, $t = \pi/2$) no optimality can be obtained. This is due to the fact that the set $\{\hat{u}(t;a) \mid a \in A\}$ (a rotating square) can not optimally be wrapped in intervals.

Figure 1. Solution of the differential system

$$(4) \qquad \begin{cases} u_1' = u_2, \\ u_2' = -u_1 \end{cases}$$

with the initial data $u(0) = A$

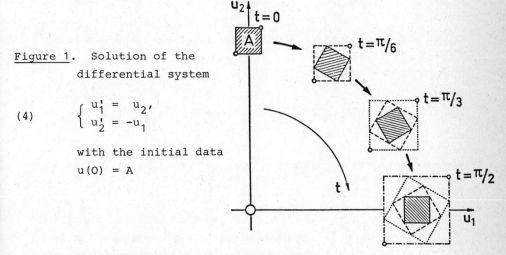

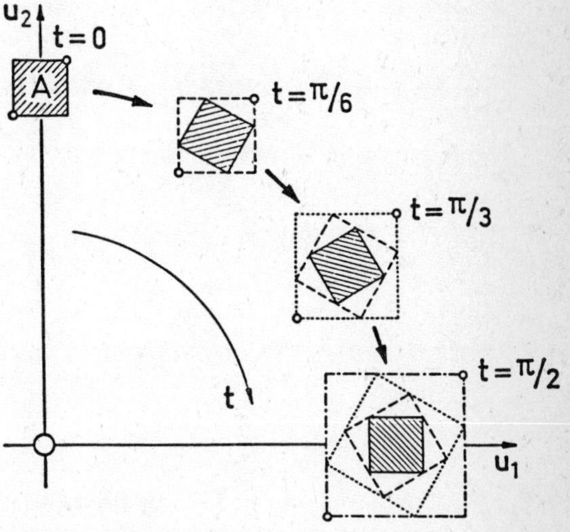


One can show rather easily that after only one revolution ($t = 2\pi$) a blow up of the optimal interval inclusion occurs by a factor of $e^{2\pi} = 535.4 \ldots$ (!). This is due to the use of intervals for U(t). Such a result is most certainly completely intolerable. It occurs, although the system (4) is extremely simple, namely

i) dimension $n = 2$,

ii) linear system,

iii) homogenous system,

iv) constant coefficients,
i.e. autonomous system.

Hence, in a more general case ($n > 2$, nonlinear) one expects the worst. Is this true? What is the reason? What can be done? In the following Sections answers to these questions will be given.

## 3. Systems without wrapping effect

Fortunately, there are large classes of differential equations, where no wrapping effect occurs. A very simple such class is given in what follows.

Definition: Let $f\colon I \times \mathbb{R}^n \to \mathbb{R}^n$ and denote $f = (f_1, f_2, \cdots, f_n)$ and $f_i = f_i(t, y_1, y_2, \cdots, y_n)$ for $i = 1(1)n$. Then f is called quasiisotone if all components $f_i$ are isotone (monotonically ascending) with respect to all variables $y_1, y_2, \cdots, y_{i-1}, y_{i+1}, \cdots, y_n$; but not necessarily to $y_i$ for $i = 1(1)n$.

Remark: For $n = 1$ any function f is quasiisotone.

The following Theorem solves Problem I for the class of (in general nonlinear) differential equations (1), where the right hand side f is quasiisotone:

Theorem: Let the function f in the equation (1) be quasiisotone and continuous and consider as sets A and U only intervals $A = [\underline{a}, \overline{a}]$ and $U(t) = [\underline{u}(t), \overline{u}(t)]$. Then no wrapping effect occurs and the inclusion (3) holds with the following weak optimality

$$\underline{u}(t, \underline{a}) \leqq \hat{u}(t, a) \leqq \overline{u}(t, \overline{a})$$

for all solutions $\hat{u}$ of (1) and (2'). Herein the functions $\underline{u}$ and $\overline{u}$ are the (existing) minimal and maximal solutions of (1), (2).

Remark: This Theorem can not be used for Moore's Example of Section 2 since the system (4) is not quasiisotone.

There is a more general Theorem which solves both Problem I and Problem II. It will not be printed here; see [5].

## 4. Linear Systems. Problem I

In this Section it will be shown that the wrapping effect can be completely explained, understood and avoided if the system (1) is linear. Hence (1) is written in the form

$$u' = Gu + h, \tag{5}$$

where $G: I \to \mathbb{R}^{n\times n}$ and $h: I \to \mathbb{R}^n$. For simplicity it is assumed that $G, h \in C(I)$. Then the problem (5), (2) has exactly one solution $\hat{u} \in C^1(I)$ which can be written as

$$\hat{u}(t;a) = X(t)a + X(t)\int_0^t X^{-1}(s)h(s)ds. \tag{6}$$

Herein the real matrix function

$$X(t) := \exp \int_0^t G(s)ds \tag{7}$$

is given with the given function G(t). It is also the uniquely determined integral basis to the homogenous system (5) under the initial condition

$$X(0) = E \quad (= \text{unit matrix}).$$

It is well known that for all $t \in I$ the inverse matrix $X^{-1}(t)$ always exists.

The dependence of the solution $\hat{u}(t;a)$ in formula (6) with respect to the initial values a is obviously linear i.e.

$$\hat{u}(t;\sigma a_1 + \tau a_2) = \sigma\hat{u}(t;a_1) + \tau\hat{u}(t;a_2)$$

$$\text{for all } \sigma,\tau \in \mathbb{R};\ a_1,a_2 \in \mathbb{R}^n.$$

Hence, the transformation (6) which maps the initial values $a \in \mathbb{R}^n$ for any fixed $t \in I$ into $\hat{u}(t;a) \in \mathbb{R}^n$ is an affine transformation. From this fact follows immediately the

Theorem: If the initial set A in (2') is a

straight line,
simplex,
parallelepiped,
ellipsoid
convex set, etc.,

then the set $\{\hat{u}(t;a) \mid a \in A\}$ belongs to the same corresponding class.

This Theorem explains immediately the negative result of Moore's Example in Section 2: An n-dimensional interval is an n-dimensional rectangle with axis-parallel sides. This will be transformed for $t > 0$ by formula (6) into an n-dimensional parallelepiped. This can, in general, be "wrapped" (included) by an interval only with a certain loss. By computing $\hat{u}(t;a)$ at many steps $0 < t_1 < t_2 < \cdots$ and wrapping it in an interval at each step this loss occurs at any time $t_1, t_2, \cdots$ and may, therefore, multiply and grow exponentially for large values of t. With this insight in the wrapping effect it is obvious to use the following

Countermeasure: For linear systems (5) do not use intervals for U(t) in the inclusion (3). Instead compute the transformation matrix X(t) as defined in (7). Then the set

(8) $$U(t) := X(t)A + X(t) \int_0^t X^{-1}(s)h(s)ds$$

is the optimal inclusion of all solutions $\hat{u}$ to the initial systems (5), (2').

This neat formula (8), unfortunately, does not say how the matrix-function X(t) should be computed. There are several possibilities. One of them is to evaluate all the eigenpairs of the matrix G(t) for each fixed value $t \in I$. From them one could then construct locally an integral basis X(t) to the homogenous system (5). Another such possibility is to integrate the matrix function G(t) and then to compute the exponential function in the formula (7) by, say, an infinite series. Both methods are, unfortunately, very laborious and time consuming.

There is, fortunately, a much easier way by exploiting the formula (8) directly: If the set A is eigher a simplex or a parallelepiped it is completely determined by n+1 corners. Hence, it is sufficient to solve (5), (2) for those corners $a_\nu$ of A for $\nu = 0(1)n$. Then the solutions $\hat{u}(t;a_\nu)$ of those n+1 real problems (5)', (2) give the corners of the desired optimal inclusion; see Figure 2 for n = 2 and an initial interval A.

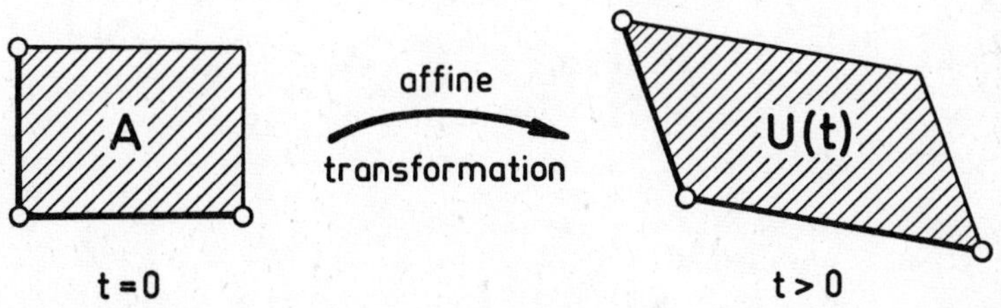


Figure 2. The transformation (8) for n = 2.

This idea has been discovered independently by R. Lohner [3] and by the author [4]. A numerical evaluation has been performed by J. Conradt [1]. It has been applied to the system (4) of Moore's Example of Section 2. His results include numerical integration, round off errors, etc. At t = 2125 (approximately 338 revolutions) his numerical results show a loss of approximately 4 decimal digits. The naive use of an interval inclusion as in Section 2 would, in contrast, have given a loss of 923 (!) decimal digits. What a difference due to the use of intelligence and formula (8) instead of a mindless naive interval computation!

## 5. Linear Systems. Problem II

In the last Section the following fact has been heavily exploited: The solution $\hat{u}(t;a)$ is a linear function with respect to the initial value a. This can be seen from formula (6). The same formula indicates in addition, that the solution $\hat{u}$ of (5), (2) is also linearly dependent on the function h(t) in (5). Hence, at first sight, it looks as if the idea of Section 4 could also be applied to the differential inclusion

(5') $\quad u' \in Gu + H.$

The difference between (5) and (5') lies in the fact that the real function h(t) has been "blown up" to a set function H(t). Assume, for simplicity, that H(t) is a bounded set in $\mathbb{R}^n$ for any $t \in I$, consisting of continuous function $h \in H$.

This hope and desire is, however, not true, unfortunately. Hence the ideas and methods of the preceding Section 4 can not be carried over to the solution of Problem II. What a pity!

The reason for this unfortunate fact is that the mapping $\int_0^t X(t)X^{-1}(s)h(s)ds$ in formula (6) is in general not an affine mapping. Hence the utilisation of affinity of Section 4 can not be used with respect to the function h in (6).

There is, however, one alternative. Define the continuous function $k: I \to \mathbb{R}^n$ by $k(t) := X^{-1}(t)h(t)$, i.e. let

$$h(t) = X(t)k(t).$$

Then the system (5) is replaced by

(9) $\quad u' = Gu + Xk$

and the initial value problem (9), (2) has the (obvious) solution (see (6))

$$\hat{u}(t) = X(t)[a + \int_0^t k(s)ds].$$

Kindly note that the matrix function X(t) is defined by formula (7). Hence

X(t) is "known" when G(t) is "given". Since X is always nonsingular the known function h defines an also known function k. Hence the two forms (5) and (9) of a linear system are both equivalent theoretically and from a computational point of view.
There are even real life problems which lead to linear equations of the type (9) instead of (5); where now the functions G,X and k are primarily given. In such a case naturally no transformation from (5) to (9) is needed.

By expanding the function k(t) into a bounded function set K(t) one gets the inclusion

(9') $u' \in Gu + XK$

with the optimal inclusion set function

(10) $U(t) := X(t)[A + \int_0^t K(s)ds]$

to all initial value inclusion problems (9'), (2').

Result:
It should be repeated that the set function U(t) as defined by formula (10) gives an optimal inclusion for both problems I and II.

Remarks:

1) To simplify the computational work the two sets A and K in formula (10) should have the same structure; e.g. they should both be intervals or both be parallelepipeds with parallel sides or ... .

2) The optimal inclusion of solutions of a differential inclusion (9') is possible, because only the function k in equation (9) is expanded into a set while the function G remains a real (matrix) function. To my knowledge no optimal inclusion is known to the solutions of an inclusion of the type (9'), where both functions K and G are set functions.

## 6. The nonlinear case

In the two previous Sections 4 and 5 the wrapping effect could be controled completely for linear systems. Hence it suggests itself to linearize nonlinear systems and then to treat them by the methods of the Sections 4 and 5.

There are infinitely many possibilities to linearize the system (1) locally by putting

$$f(t,y) = G(t)y + X(t)k(t,y)$$

with a suitable matrix function G(t) and X(t) correspondingly defined by (7). A possible choice of G is

$$G(t) := \frac{\partial f}{\partial y}(t,\tilde{u}(t))$$

with a suitable approximation $\tilde{u}$ to a solution $\hat{u}$ of (1), (2). If f in (1) is continuous then G and k may also be chosen as continuous functions. In this case the initial value problem

$$u'(t) = G(t)u(t) + X(t)k(t,u(t))$$

with the initial condition (2) is obviously equivalent to the Volterra integral equation

$$u(t) = X(t)[a + \int_0^t k(s,u(s))ds].$$

All solutions $\hat{u}$ of this equation can be bounded as in (3) by suitable set functions U and these bounds U can be found and evaluated by standard methods of interval mathematics. These methods also apply if the initial value a and the function k(t,y) are replaced by sets A and K(t,y).

It is not to be expected that in the nonlinear case the wrapping effect can be eliminated completely with this technique. One can expect, however, that the remaining wrapping effect will be "small" if the right hand side f of (1) is only "mildly" nonlinear. If, opposite to this, the function f(t,y) is "strongly" nonlinear with respect to y <u>no</u> method is known to the author to describe and to ban the occuring "nonlinear wrapping effects".

## 7. Final remarks

The discusssion on the "wrapping effect" of interval methods for the numerical solution of differential equations goes on and on since more than 20 years. Nevertheless one should never lose sight of the following essential facts:

The customary real numerical methods give approximations, not solutions! The error of them is in general not known. It may be very large in particular cases, unknown to the user of such methods.

Compared with this the interval methods have the big advantage to give always guaranteed bounds to the (normally unknown) solutions. Even if they are unfavorable (which may occur with the naive use of interval arithmetic; or without consideration of the wrapping effect) they do give an exact information.

*Summary: Any errorbound - even a pessimistic one - is better than no information at all about the error of an approximation. The responsibility of interval research is to find favorable error bounds.*

## References


[1] Conradt, Jürgen: Ein Intervallverfahren zur Einschließung des Fehlers einer Näherungslösung bei Anfangswertaufgaben für Systeme von gewöhnlichen Differentialgleichungen. Diplomarbeit. Freiburger Intervall-Berichte 80/1. Institut für Angewandte Mathematik, Universität Freiburg i.Br. (1980).

[2] Gambill, Thomas N. and Robert D. Skeel: Logarithmic Reduction of the Wrapping Effect with Applications to Ordinary Differential Equations. University of Illinois, Manuscript (1984).

[3] Lohner, Rudolf: Anfangswertaufgaben im $\mathbb{R}^n$ mit kompakten Mengen für Anfangswerte und Parameter. Diplomarbeit am Institut für Angewandte Mathematik, Universität Karlsruhe (1978).

[4] Nickel, Karl: Bounds for the Set of Solutions of Functional-Differential Equations. MRC Technical Summary Report # 1782, University of Wisconsin, Madison (1977). Annales Polonici Mathematici 42 (1983), 241-257.

[5] Nickel, Karl: Ein Zusammenhang zwischen Aufgaben monotoner Art und Intervall-Mathematik. Numerical Treatment of Differential Equations, Proc. of a conf. held at Oberwolfach, July 4-10, 1976. Ed. by R. Bulirsch, R.D. Grigorieff, and J. Schröder, Springer Verlag, Berlin, Heidelberg, New York, 121-132 (1978).

[6] Nickel, Karl: Using Interval Methods for the Numerical Solution of ODE's. MRC Technical Summary Report # 2590. University of Wisconsin, Madison (1983). Freiburger Intervall-Berichte 83/10. Institut für Angewandte Mathematik, Universität Freiburg i.Br., 13-44 (1983). To appear in ZAMM.

# ARITHMETIC OF CIRCULAR RINGS

MIODRAG S. PETKOVIĆ
*Faculty of Electronic Engineering, Niš, Yugoslavia*

ŽARKO M. MITROVIĆ
*Technical Faculty, Zrenjanin, Yugoslavia*

LJILJANA D. PETKOVIĆ
*Faculty of Mechanical Engineering, Niš, Yugoslavia*

## 1. INTRODUCTION

Circular arithmetic, introduced by I. Gargantini and P. Henrici [2] as an extension of the complex arithmetic, provided the formulation of methods for solving some problems of computational complex analysis (e.g. the inclusion of the polynomial complex zeros [2],[3], circular approximation of the closed regions in the complex plane [1],[4], [6],[8], the evaluations of complex functions over a disk as an argument [4],[5],[6],[8],[9], etc.). Applying these methods, sometimes a problem of evaluation with the disks which contain the origin arises (for example, inversion of a disk Z, evaluation of the complex functions $z \mapsto \ln z$, $z \mapsto z^{1/k}$ over a disk Z, where $0 \in Z$). In some cases, this problem can be overcome by evaluation with the annulus $\{z : r \leqq |z-c| \leqq R\}$ instead of the disk $\{z : |z-c| \leqq R\}$ if the origin is contained in the internal disk $\{z : |z-c| \leqq r\}$. In this way, an isolation of zero has been done. Besides, in some cases, a set of points $\Omega$ in the complex plane (e.g. an opened or closed curve which is characterized by the circularity) can be enclosed by an annulus, say Z. For a given analytic function f, this inclusion enables to consider in the sequel an *annular* function F such that

$$F(Z) \supseteq f(Z) = \{f(z) : z \in Z\} \supseteq f(\Omega) \qquad (\Omega \subseteq Z).$$

Note that T.J. Rivlin [7] used the annulli for assignation of a measure of circularity to a given compact set in the plane. His approach is to determine the best annulus which contains the given set according to the size of the annulus.

In this paper we shall pay attention only to the problem of defining the basic arithmetic operations with the circular rings. Further, we shall point out some possibilities to use the arithmetic of annulli, which is an extension of circular arithmetic. Note that this generalization carries some disadventages, for example, more complicated arithmetic operations and reduction of these operations to the operations in circular arithmetic in some cases.

## 2. ARITHMETIC OPERATIONS

A set $Z=\{z : r \leq |z-c| \leq R,\ c \in C,\ 0 \leq r \leq R\}$, denoted by $Z=\{c; (r,R)\}$, will be called a *circular ring* or *annulus*. Here $c$ is the center, $r$ and $R$ are internal and external radii of the annulus $Z$, respectively. The set of all circular rings will be denoted by $A(C)$. In the special case, for $r=0$, the annulus $Z = \{c\ ;\ (0,R)\}$ reduces to the disk $Z=\{z: |z-c| \leq R\}$, denoted shorter by $Z = \{c\ ; R\}$.

Sometimes, an annulus whose center is at the origin, will be marked by $\overset{\circ}{Z}$, that is

$$\overset{\circ}{Z} = \{0\ ;\ (r,R)\} = \{\rho e^{i\phi}: r \leq \rho \leq R,\ 0 \leq \phi \leq 2\pi\}.$$

An annulus $Z = \{c\ ;\ (r,R)\}$ can be presented in the form

$$Z = \{c\ ;\ (r,R)\} = c + \overset{\circ}{Z}\ , \tag{1}$$

because of

$$\begin{aligned} Z &= \{c + \rho e^{i\phi}: r \leq \rho \leq R,\ 0 \leq \phi \leq 2\pi\} \\ &= \{c\} + \{\rho e^{i\phi}: r \leq \rho \leq R,\ 0 \leq \phi \leq 2\pi\}. \end{aligned}$$

*ADDITION OF A SCALAR AND AN ANNULUS :*

Since

$$w + \{c\ ;\ (r,R)\} = \{w + z: r \leq |z-c| \leq R\} = \{z: r \leq |z-(w+c)| \leq R\},$$

we have

$$w + \{c\ ;\ (r,R)\} = \{w + c\ ;\ (r,R)\}. \tag{2}$$

*MULTIPLICATION OF A SCALAR AND AN ANNULUS :*

Let $w \in C$, $\overset{\circ}{Z} = \{0\ ;\ (r,R)\}$ and $Z = \{c\ ;\ (r,R)\}$. First, we have

$$\begin{aligned} w\cdot\overset{\circ}{Z} = w\cdot\{0\ ;\ (r,R)\} &= w\cdot\{z = \rho e^{i\phi}: r \leq \rho \leq R,\ 0 \leq \phi \leq 2\pi\} \\ &= \{z' = |w|\rho e^{i(\phi+\arg w)}: r \leq \rho \leq R,\ 0 \leq \phi \leq 2\pi\} \\ &= \{z' = \rho' e^{i\phi'}: |w|r \leq \rho' \leq |w|R,\ 0 \leq \phi' \leq 2\pi\}, \end{aligned}$$

wherefrom

$$w\cdot\overset{\circ}{Z} = \{0\ ;\ (|w|r, |w|R)\}. \tag{3}$$

On the basis of (2) and (3) we find

$$\begin{aligned} w\cdot Z = w\cdot\{c\ ;\ (r,R)\} &= w\cdot(c + \overset{\circ}{Z}) = wc + w\cdot\overset{\circ}{Z} \\ &= wc + \{0\ ;\ (|w|r, |w|R)\} = \{wc\ ;\ (|w|r, |w|R)\}, \end{aligned}$$

that is,

$$w\cdot\{c\ ;\ (r,R)\} = \{wc\ ;\ (|w|r,|w|R)\}. \tag{4}$$

*INCLUSION AND DISJUNCTION OF THE ANNULLI:*

Let $Z_i = \{c_i\ ;(r_i,R_i)\}\in A(C)$, $i=1,2$. The annulus $Z_2$ *contains* the annulus $Z_1$, denoted by $Z_1\subseteq Z_2$, if and only if

$$\Big((R_2>R_1)\wedge(r_1>r_2)\wedge(|c_2-c_1|<\min\{r_1-r_2,R_2-R_1\})\Big)\vee\Big(R_2-R_1>|c_1|>r_2+R_1\Big).$$

The annulli $Z_1$ and $Z_2$ are *disjoint* (i.e. $Z_1\cap Z_2=\emptyset$) if and only if one of the following conditions is valid:

$$\Big(|c_2-c_1|>R_1+R_2\Big)\vee\Big((r_2>R_1)\wedge(|c_2-c_1|<r_2-R_1)\Big)\vee\Big((r_1>R_2)\wedge(|c_2-c_1|<r_1-R_2)\Big).$$

*ADDITION AND SUBSTRACTION:*

Let $\overset{o}{Z}_i = \{0\ ;\ (r_i,R_i)\}$, $i=1,2$. Then

$$\overset{o}{Z}_1+\overset{o}{Z}_2 = \{\rho_1e^{i\phi_1}+\rho_2e^{i\phi_2} : r_i\leq\rho_i\leq R_i,\ 0\leq\phi_i\leq 2\pi,\ i=1,2\}$$

$$= \{0\ ;\ (r,R)\},$$

where

$$r = \min_{\substack{\rho_1,\rho_2\\ \phi_1,\phi_2}} |\rho_1e^{i\phi_1}+\rho_2e^{i\phi_2}|,$$

$$R = \max_{\substack{\rho_1,\rho_2\\ \phi_1,\phi_2}} |\rho_1e^{i\phi_1}+\rho_2e^{i\phi_2}|.$$

It is sufficient to take $\phi_i\in[0,\pi]$ $(i=1,2)$. Obviously,

$$R = \max_{r_1\leq\rho_1\leq R_1}\rho_1 + \max_{r_2\leq\rho_2\leq R_2}\rho_2 = R_1+R_2,$$

$$r = \min_{\substack{r_1\leq\rho_1\leq R_1\\ r_2\leq\rho_2\leq R_2}} |\rho_1-\rho_2|.$$

The value of r is given by

$$r = \begin{cases} r_2-R_1, & \text{if } r_2>R_1,\\ r_1-R_2, & \text{if } r_1>R_2,\\ 0, & \text{otherwise, that is, if } \overset{o}{Z}_1\cap\overset{o}{Z}_2\neq\emptyset. \end{cases} \tag{5}$$

Now, we have

$$\{0 ; (r_1,R_2)\} + \{0 ; (r_2,R_2)\} = \{0 ; (r,R_1+R_2)\}, \tag{6}$$

where r is defined by (5).

According to (6) it follows

$$Z_1 + Z_2 = \{c_1 ; (r_1,R_1)\} + \{c_2 ; (r_2,R_2)\}$$
$$= (c_1 + \overset{\circ}{Z}_1) + (c_2 + \overset{\circ}{Z}_2) = c_1 + c_2 + (\overset{\circ}{Z}_1 + \overset{\circ}{Z}_2),$$

or

$$Z_1 + Z_2 := \{c_1 + c_2 ; (r,R_1+R_2)\}. \tag{7}$$

Thus, *the addition* in the set A(C) is defined by (7), where r is given by (5).

It is easy to prove that for the addition of the annulli the following is valid:

$$Z_1 + Z_2 = Z_2 + Z_1 \qquad (\textit{commutativity}),$$
$$(Z_1 + Z_2) + Z_3 = Z_1 + (Z_2 + Z_3) \qquad (\textit{associativity}).$$

Let $Z_i = \{c_i ; (r_i,R_i)\}$ $(i=1,2,3)$. The final result of the addition of three annulli we denote by $Z_1 + Z_2 + Z_3$, and it is given by

$$Z_1 + Z_2 + Z_3 = \{c_1 + c_2 + c_3 ; (r,R_1 + R_2 + R_3)\}, \tag{8}$$

where

$$r = \begin{cases} r_1 - R_2 - R_3, & \text{if } r_1 > R_2 + R_3, \\ r_2 - R_1 - R_3, & \text{if } r_2 > R_1 + R_3, \\ r_3 - R_1 - R_2, & \text{if } r_3 > R_1 + R_2, \\ 0, & \text{otherwise.} \end{cases}$$

*The substraction* of the annulli is defined using (3) and (7):

$$\{c_1 ; (r_1,R_1)\} - \{c_2 ; (r_2,R_2)\} := \{c_1 ; (r_1,R_1)\} + \{-c_2 ; (r_2,R_2)\}$$
$$= \{c_1 - c_2 ; (r,R_1 + R_2)\}, \tag{9}$$

where, again, r is given by (5).

*MULTIPLICATION:*

The product of two annulli $Z_1Z_2$ is not an annulus in general. For this reason, an extended set $Z_1 \otimes Z_2$ in the form of an annulus, such that $Z_1 \otimes Z_2 \supseteq Z_1Z_2$, is introduced. The symbol $\otimes$ denotes the multiplication in the set A(C).

The product of the annulli $\{0;(r_1,R_1)\}$ and $\{0;(r_2,R_2)\}$ is given by

$$\{0;(r_1,R_1)\} \otimes \{0;(r_2,R_2)\} = \{\rho_1 e^{i\phi_1}: r_1 \leqq \rho_1 \leqq R_1,\ 0 \leqq \phi_1 \leqq 2\pi\} \otimes$$
$$\{\rho_2 e^{i\phi_2}: r_2 \leqq \phi_2 \leqq R_2,\ 0 \leqq \phi_2 \leqq 2\pi\}$$
$$= \{\rho e^{i\phi}: r_1 r_2 \leqq \rho \leqq R_1 R_2,\ 0 \leqq \phi \leqq 2\pi\}$$
$$= \{0\ ;\ (r_1 r_2, R_1 R_2)\},$$

where $\rho = \rho_1\rho_2$ and $\phi = \phi_1+\phi_2$. Accordingly,

$$\{0\ ;\ (r_1,R_1)\} \otimes \{0\ ;\ (r_2,R_2)\} = \{0\ ;\ (r_1 r_2, R_1 R_2)\}. \tag{10}$$

It is obvious that

$$\mathring{Z}_1 \otimes \mathring{Z}_2 = \mathring{Z}_1 \mathring{Z}_2 = \{z_1 z_2 : z_1 \in \mathring{Z}_1, z_2 \in \mathring{Z}_2\}.$$

By induction we show that

$$\prod_{k=1}^{n} \mathring{Z}_k = \left\{ 0\ ;\ \left( \prod_{k=1}^{n} r_k\ ,\ \prod_{k=1}^{n} R_k \right) \right\}$$

is valid, wherefrom, in a special case, we have

$$\{0\ ;\ (r,R)\}^n = \{0\ ;\ (r^n, R^n)\}\ . \tag{11}$$

Let $Z_i = \{c_i\ ;\ (r_i,R_i)\} \in A(C)$ $(i=1,2)$. According to (10) we find

$$Z_1 Z_2 = \{c_1\ ;\ (r_1,R_1)\}\cdot\{c_2\ ;\ (r_2,R_2)\}$$
$$= (c_1 + \{0;(r_1,R_1)\})\cdot(c_2 + \{0;(r_2,R_2)\})$$
$$\subseteq c_1 c_2 + c_2\cdot\{0;(r_1,R_1)\} + c_1\cdot\{0;(r_2,R_2)\} +$$
$$\{0;(r_1,R_1)\}\cdot\{0;(r_2,R_2)\}.$$

Applying (4) and (10), we obtain that

$$Z_1 Z_2 \subseteq c_1 c_2 + \{0;(|c_2|r_1, |c_2|R_1)\} + \{0;(|c_1|r_2, |c_1|R_2)\} +$$
$$\{0;(r_1 r_2, R_1 R_2)\}.$$

In view of (8) and (9), it follows

$$Z_1 Z_2 \subseteq c_1 c_2 + \{0\ ;\ (r,R)\} = \{c_1 c_2\ ;\ (r,R)\},$$

where

$$R = |c_2|R_1 + |c_1|R_2 + R_1 R_2 \tag{12}$$

and

$$r = \begin{cases} r_1r_2 - |c_2|R_1 - |c_1|R_2, & \text{if } r_1r_2 > |c_2|R_1 + |c_1|R_2 \\ |c_1|r_2 - |c_2|R_1 - R_1R_2, & \text{if } |c_1|r_2 > |c_2|R_1 + R_1R_2 \\ |c_2|r_1 - |c_1|R_2 - R_1R_2, & \text{if } |c_2|r_1 > |c_1|R_2 + R_1R_2 \\ 0, & \text{otherwise.} \end{cases} \tag{13}$$

For the product $Z_1 \otimes Z_2$ of the annulli $Z_1$ and $Z_2$ we shall adopt the extended set $\{c_1c_2\,;\,(r,R)\}$, that is

$$Z_1Z_2 \subseteq Z_1 \otimes Z_2 := \{c_1c_2\,;\,(r,R)\}, \tag{14}$$

where r and R are given by (12) and (13) respectively.

It is easy to show that four cases in (13) are disjoint. Let us show, for example, that the inequalities

$$|c_1|r_2 > |c_2|R_1 + R_1R_2,$$
$$|c_2|r_1 > |c_1|R_2 + R_1R_2,$$

can not be valid simultaneously. Rewrite the above inequalities in the form

$$y = |c_1|r_2 - |c_2|R_1 > R_1R_2, \tag{*}$$
$$|c_2|r_1 - |c_1|R_2 > R_1R_2, \tag{**}$$

and put

$$R_1 = r_1 + \varepsilon_1, \quad R_2 = r_2 + \varepsilon_2 \quad (\varepsilon_1, \varepsilon_2 > 0).$$

Suppose that (*) holds. Then $y > 0$, while the left-hand side of (**) becomes

$$\begin{aligned} |c_2|r_1 - |c_1|R_2 &= |c_2|(R_1-\varepsilon_1) - |c_1|(r_2+\varepsilon_2) \\ &= |c_2|R_1 - |c_2|\varepsilon_1 - |c_1|r_2 - |c_1|\varepsilon_2 \\ &= -y - |c_2|\varepsilon_1 - |c_1|\varepsilon_2 < 0. \end{aligned}$$

Thus, if (*) is valid, then (**) does not hold.

Note that, in the case when $Z_1$ and $Z_2$ are disks, that is, $Z_1 = \{c_1\,;\,(0,R_1)\} = \{c_1\,;\,R_1\}$ and $Z_2 = \{c_2\,;\,(0,R_2)\} = \{c_2\,;\,R_2\}$, the definition (14) for the product of annulli reduces to the definition for the product of disks introduced by I. Gargantini and P. Henrici [2].

Since the exchange of the indices at (12) and (13) (i.e. $1 \to 2$ and $2 \to 1$) does not cause any modification in the expressions for R and r, we conclude that

$$Z_1 \otimes Z_2 = Z_2 \otimes Z_1,$$

i.e., the product of the annulli, introduced by (14), is *commutative*.

*INVERSION OF ANNULUS:*

Defining the inversion of an annulus, we shall use the formula for the inversion of a disk $\{c\,;r\}$ which does not contain the origin,

$$\{c\,;r\}^{-1} = \left\{\frac{\bar{c}}{|c|^2 - r^2}\,;\,\frac{r}{|c|^2 - r^2}\right\}. \tag{15}$$

An inverse annulus $Z^{-1}$, where $Z = \{c;(r,R)\}$, will be defined in the case when $0 \notin Z$, i.e. if $|c| > R$ or $|c| < r$ holds (which is equivalent to $0 \notin \{c\,;R\}$ and $0 \in \{c\,;r\}$, respectively). Defining the inverse annulus we shall use the inverse disks

$$\{c\,;R\}^{-1} = \{w_1\,;\rho_1\}, \qquad \{c\,;r\}^{-1} = \{w_2\,;\rho_2\},$$

where, on the basis of (15),

$$w_1 = \frac{\bar{c}}{|c|^2 - R^2}, \quad \rho_1 = \frac{R}{|c|^2 - R^2}, \quad w_2 = \frac{\bar{c}}{|c|^2 - r^2}, \quad \rho_2 = \frac{r}{|c|^2 - r^2}.$$

We observe that the centers $w_1$ and $w_2$ are not overlapping, which means that the circles $\Gamma_1 = \{w\colon |w-w_1| = \rho_1\}$ and $\Gamma_2 = \{w\colon |w-w_2| = \rho_2\}$ are not concentric (see Fig. 1 and 2). For this reason, an extension of the exact set $\{z^{-1}\colon r \leqq |z-c| \leqq R\}$ must be taken to be the inverse annulus $Z^{-1}$. The extension has to be performed so that the implication

$$0 \notin Z \Rightarrow 0 \notin Z^{-1}$$

holds, which is of the essential interest in executing the arithmetic operations in the set $A(C)$. We shall distinguish two cases: $0 \notin \{c;R\}$ and $0 \in \{c;r\}$.

*a)* $0 \notin \{c\,;R\}$

In this case the origin is outside of the external circle and, thus, $|c| > R$ is valid. The circumferences $\Gamma_1$ and $\Gamma_2$ are not concentric so that we shall construct the inverse annulus $Z^{-1}$ by extending the exact range $\{z^{-1}\colon r \leqq |z-c| \leqq R\}$ over the internal disk $\{c\,;r\}$ taking the point $w_1$ for the center of the inverse annulus $Z^{-1}$. In this manner the implication $0 \notin Z \Rightarrow 0 \notin Z^{-1}$ is provided (because the condition $0 \notin \{c\,;R\}$ provides that $0 \notin \{c\,;R\}^{-1}$).

From Fig. 1 we have

$$Z^{-1} := \{w_1\,;\,(\max\{0\,,\rho_2 - |w_1-w_2|\}\,,\rho_1)$$

$$= \frac{1}{|c|^2 - R^2}\left\{\bar{c}\,;\left(\max\left\{0\,,\frac{r|c| - R^2}{|c| - r}\right\},\,R\right)\right\}.$$

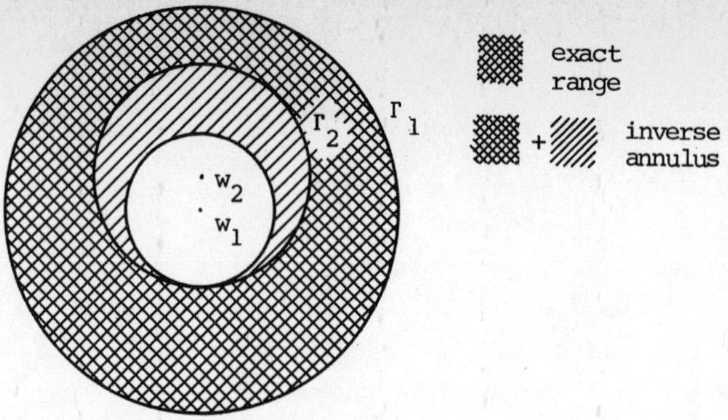


Fig. 1 Inverse annulus: the case $0 \notin \{c\,;R\}$

*b)* $0 \in \{c\,;r\}$

In this case we have $|c| < r$ and

$$\{c\,;R\}^{-1} = \{w\colon |w-w_1| \geqq \rho_1\}, \quad \{c\,;r\}^{-1} = \{w\colon |w-w_2| \geqq \rho_2\},$$

where

$$w_1 = \frac{\bar{c}}{|c|^2 - R^2}, \quad \rho_1 = \frac{R}{R^2 - |c|^2}, \quad w_2 = \frac{\bar{c}}{|c|^2 - r^2}, \quad \rho_2 = \frac{r}{r^2 - |c|^2}.$$

The origin belongs to the interior of the smaller circle $\Gamma_1 = \{w\colon |w - w_1| = \rho_1\}$ because of $|w_1| < \rho_1$. For this reason the extension of the exact range will be performed outside of the circle $\Gamma_2$, taking the point $w_1$ for the center of inverse annulus.

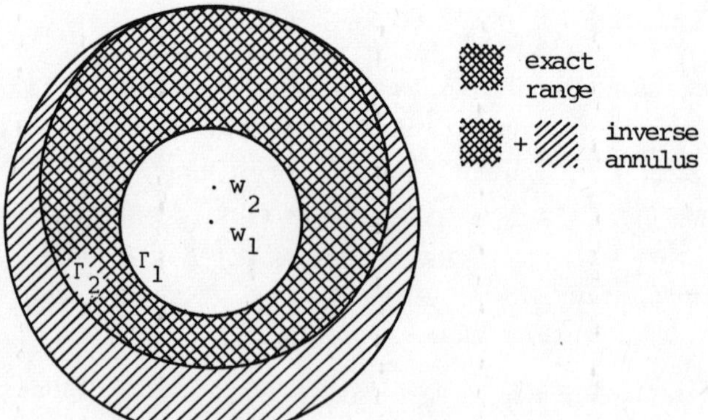


Fig. 2 Inverse annulus: the case $0 \in \{c\,;r\}$

From Fig. 2 we find

$$Z^{-1}: = \{w_1 \ ; \ (\rho_1 \ , |w_2 - w_1| + \rho_2)\}$$
$$= \frac{1}{R^2 - |c|^2}\left\{-\bar{c} \ ; \left(R \ , \frac{R^2 - r|c|}{r - |c|}\right)\right\}.$$

Thus, the inverse annulus $Z^{-1}$ $(0 \notin Z)$ is defined by

$$Z^{-1}: = \begin{cases} \dfrac{1}{|c|^2 - R^2}\left\{\bar{c} \ ; \left(\max\left\{0 \ , \dfrac{r|c| - R^2}{|c| - r}\right\} , R\right)\right\} & \text{if } |c| > R \ , \\ \dfrac{1}{R^2 - |c|^2}\left\{-\bar{c} \ ; \left(R \ , \dfrac{R^2 - r|c|}{r - |c|}\right)\right\} & \text{if } |c| < r \ . \end{cases} \tag{16}$$

*DIVISION:*

Let the symbol $\odot$ denotes the division in A(C). Using the definitions (14) and (16) for the multiplication and the inversion, we define the operation of division in A(C) as follows:

$$Z_1 \odot Z_2: = Z_1 \otimes Z_2^{-1} \quad (0 \notin Z_2). \tag{17}$$

## 3. ANNULAR FUNCTIONS

Let f be an analytic function defined on the region D in the complex plane, and let Z be an annulus contained in D. The set $f(Z) = \{f(z): z \in Z\}$ is not an annulus in general. In order to use the arithmetic of annulli, a necessity arises for introducing an annular function $F: G \to H$ $(G,H \subset A(D))$ such that the following is valid:

$$\begin{aligned} F(Z) &\supseteq f(Z) \quad \text{for all } Z \in A(D), \\ F(z) &= f(z) \quad \text{for all } z \in Z. \end{aligned} \tag{18}$$

The function F such that (18) is satisfied is called an *annular inclusive extension* of f.

Let $\Omega$ be a set of points in the complex plane such that it can be suitably „enclosed" by an annulus Z, that is, $\Omega \subseteq Z$. Assume that f is a complex function such that the set f(Z) is a closed region and let F be an annular inclusive extension of f. Since $f(\Omega) \subseteq f(Z) \subseteq F(Z)$, we can consider the annulus F(Z) (for a given Z), which presents an annular approximation of the set $f(\Omega)$, instead of the image $f(\Omega)$. The approximation („covering") of the set $\Omega$ by an annulus Z is of special interest if the set $\Omega$ is characterized by a certain circularity. $\Omega$ can be also some opened or closed curve, $\Omega = \{\omega(t): \omega(t) = u(t) + i\,v(t) \ , \ t \in (\alpha,\beta)\}$, as it has been mentioned in the beginning. For example, an ellipse given by $\omega(t) = a\cos t + i\,b\sin t$ $(t \in [0,2\pi),\ a > b)$, can be bounded by an annulus $\overset{\circ}{Z} = \{0\ ;\ (b,a)\}$. This approximation is better if the quotient $\frac{a}{b}$ is closer to 1.

## REFERENCES


1. BOERSKEN, N.C.: *Komplexe Kreis-Standardfunktionen*. Freiburger Intervall-Berichte 2 (1978), 1-102.
2. GARGANTINI, I., HENRICI, P.: *Circular arithmetic and the determination of polynomial zeros*. Numer. Math. 18 (1972), 305-320.
3. HENRICI, P.: Applied and Computational Complex Analysis, Vol. 1. John Wiley and Sons, New York-London-Sydney 1974.
4. PETKOVIĆ, L.D.: On some approximations by disks (in Serbo-Croatian). Ph. D. Thesis. University of Kragujevac 1985.
5. PETKOVIĆ, L.D.: *A note on the evaluation in circular arithmetic*. (to appear in ZAMM).
6. PETKOVIĆ, L.D., PETKOVIĆ, M.S.: *The representation of complex circular functions using Taylor series*. ZAMM 61 (1981), 661-662.
7. RIVLIN, T.J.: *Approximation by circles*. Computing 21 (1979), 93-104.
8. ROKNE, J., WU, T.: *The circular complex centered form*. Computing 28 (1982), 17-30.
9. WU, T.: Circular complex centered form. M. Sc. Thesis. University of Calgary 1981.

Improved Interval Bounds for Ranges of Functions

L. B. Rall

Mathematics Research Center
University of Wisconsin-Madison
610 Walnut Street

Madison, Wisconsin 53705/USA

1. Ranges of functions. The range of a real function $f:D \subset \mathbf{R} \to \mathbf{R}$ on a set $X \subset D$ is

$$R(f;X) = \{f(x) \mid x \in X\}. \tag{1.1}$$

In case $X = [a,b]$ is a closed, bounded interval and f is continuous, then $R(f;X)$ will also be an interval of the same kind. Closed, bounded intervals will be referred to simply as intervals, and the set of such intervals will be denoted by **IR**.

A fundamental problem of interval analysis is the calculation of $R(f;X)$ or at least a good approximation to it. If f is defined in terms of arithmetic operations and functions with known interval extensions, then straightforward use of interval computation gives an interval extension F of f such that

$$R(f;X) \subset F(X) \tag{1.2}$$

for $X \subset D$. This calculation has the advantage of being completely automatic, and does not require knowledge of special properties of f. Unfortunately, $F(X)$ can be such a gross overestimation of $R(f;X)$ in certain cases that it is useless for practical purposes. Furthermore, the quality of $F(X)$ as an approximation to $R(f;X)$ is generally unknown.

A number of methods have been developed for obtaining better approximations to $R(f;X)$, starting with the work of Moore [1]. The recent book by Ratschek and Rokne [5] describes a number of these techniques, and gives a substantial bibliography. Most of the approaches to this problem are based on transformation of F, usually into centered or mean-value forms [1], [5]. The method given in this paper applied to continuously differentiable functions f, and makes use of information about the monotonicity of f obtained by the process of automatic differentiation [2].

2. Monotone functions. If the function f is nondecreasing on X, then R(f;X) is simply

$$R(f;X) = [f(a), f(b)]. \tag{2.1}$$

Similarly, if f is nonincreasing on X, then

$$R(f;X) = [f(b), f(a)]. \tag{2.2}$$

Thus, the range of monotone functions can be determined by calculating only two function values. In actual practice, of course, downward rounding of the lower endpoint and upward rounding of the upper endpoint gives an interval inclusion of R(f;X) which is slightly wider than the exact range. For the time being, it will be assumed that function values are computed exactly.

A sufficient condition for (2.1) to hold for differentiable f is that

$$f'(x) \geq 0, \quad a \leq x \leq b, \tag{2.3}$$

and similarly (2.2) holds if $f'(x) \leq 0$ on X. Furthermore, suppose that f is continuously differentiable, and F' denotes an interval extension of f' obtained by interval computation. If $F'(X) \geq 0$ ($F'(X) \leq 0$), it follows that f is nondecreasing (nonincreasing) on X, and R(f;X) can be calculated directly by (2.1) or (2.2), respectively.

The additional information about the derivative of f needed above can also be obtained automatically. The values of F(X) and F'(X) can be computed by using interval differentiation arithmetic, as described below. All that is required is a formula or subroutine for f; no symbolic differentiation is necessary. If necessary, a bisection procedure can be applied to the interval X to find subintervals on which f can be guaranteed to be monotone. The resulting algorithm provides either the exact value of R(f;X), or else an inclusion of R(f;X) which is better in general than F(X).

3. Real differentiation arithmetic. It is convenient to define interval differentiation arithmetic as an extension of real differentiation arithmetic. This arithmetic can be used to calculate the values of functions and their derivatives automatically, without symbolics or numerical approximations [4]. Like interval arithmetic, real differentiation arithmetic is an ordered-pair arithmetic, with elements $U = (u,u')$, $V = (v,v')$, $\ldots \in \mathbf{R}^2$. The rules for this arithmetic are:

$$U + V = (u,u') + (v,v') = (u + v, u' + v'), \tag{3.1}$$

$$U - V = (u,u') - (v,v') = (u - v, u' - v'), \tag{3.2}$$

$$U \cdot V = (u,u') \cdot (v,v') = (u \cdot v,\ u \cdot v' + v \cdot u'), \tag{3.3}$$

$$U/V = (u,u')/(v,v') = (u/v,\ (u' - (u/v) \cdot v')/v), \quad v \neq 0. \tag{3.4}$$

The arithmetic defined in this way forms a division ring with identity, and will be denoted by **D**. If the first element of each operand pair is interpreted as a function value, and the second as a derivative value, then the first element of the result corresponds to the evaluation of the operation, and the second to the evaluation of its derivative, according to the well-known rules of calculus. If real numbers c are identified with the pairs (c,0), then it follows from the chain rule of calculus that

$$f((x,1)) = (f(x),\ f'(x)), \tag{3.5}$$

that is, the rules of differentiation arithmetic will automatically give both the value and the value of the derivative of a rational function f. More generally, the chain rule gives

$$f((u,u')) = (f(u),\ u' \cdot f'(u)), \tag{3.6}$$

which allows the definition of standard functions in D, for example,

$$e^U = e^{(u,u')} = (e^u,\ u' \cdot e^u), \tag{3.7}$$

$$\ln U = \ln(u,u') = (\ln u,\ u'/u), \tag{3.8}$$

and so on. The combination of arithmetic operations and standard functions will be called a computational system for differentiation arithmetic. It is simple to program such a computational system, particulary in a language such as Pascal-SC, which permits definition of operators and functions for various data types [3].

4. Interval differentiation arithmetic. Interval differentiation arithmetic is defined by the same rules as real differentiation arithmetic, starting with pairs of intervals instead of real numbers, and using interval arithmetic instead of real arithmetic inside the parentheses on the right sides of (3.1)-(3.3). With interval extensions of standard functions, the definitions (3.7), (3.8) and so on are used to construct a computational system for interval differentiation arithmetic. Once again, such a system is easy to program in Pascal-SC, which supports interval arithmetic as well as operator and function definitions for various data types [3].

The analog to (3.5) in interval differentiation arithmetic is

$$F((X,[1,1])) = (F(X),\ F'(X)). \tag{4.1}$$

Thus, by a direct evaluation process in this arithmetic, interval inclusions $F(X)$ of $R(f;X)$ and $F'(X)$ of $R(f';X)$ can both be obtained automatically. Here, even if $F'(X)$ is a crude approximation to the range of $f'$ on $X$, the conditions $F'(X) \geq 0$ or $F'(X) \leq 0$ are sufficient to guarantee the monotonicity of $f$, and if $f$ is monotone, then its range can be calculated exactly by (2.1) or (2.2). This observation is the basis of the algorithm described in the next section.

5. An algorithm for range calculation. Of course, if the calculation of $F'(X)$ shows that $f$ is monotone on the entire interval $X$, then $R(f;X)$ can be calculated at once. Otherwise, $X$ will be partitioned into subinterval, and either $R(f;X)$ or an approximation to it will be constructed. Let a given list of $n$ subintervals of $X$ be denoted by $L_n = \{X_1, X_2, \ldots, X_n\}$, and suppose that $R \subset R(f;X)$ is known. On each subinterval $X_i$, either $F(X_i) \subset R$, in which case $R(f;X_i)$ makes no additional contribution to $R(f;X)$, or $f$ is monotone, in which case its range can be computed directly and $R$ updated, or else $0$ is an interior point of $F'(X_i)$, in which case $X_i$ may contain a critical point of $f$. In the latter case, $X_i$ can be bisected and the resulting subintervals put on a new list for further examination. In order for the algorithm to terminate in a finite number of steps, a lower bound $\delta$ is put on the widths of the subintervals to be considered, and an upper bound $N$ is placed on the number of subintervals to be saved for further examination. For convenience, if $Y,Z$ are intervals, then $Y \mathbin{++} Z$ will denote the interval hull of $Y$ and $Z$, that is, the smallest interval which contains both $Y$ and $Z$.

The algorithm consists of the following steps:

1°. (Initialization) Take $X_1 := X$, $L_1 := \{X_1\}$, $R := [f(x),f(x)]$, where $x$ is some point in $X$.

2°. (Iteration) For $i = 1,\ldots,n$, compute $(F(X_i), F'(X_i))$

(a) If $F(X_i) \subset R$, then discard $X_i$.

(b) If $F'(X_i) \geq 0$ or $F'(X_I) \leq 0$, then compute $R := R \mathbin{++} R(f,X_i)$ and discard $X_i$.

(c) Otherwise, retain $X_i$.

3°. (Termination or continuation) Denote the list of retained intervals by $L_r$.

(a) If $L_r$ is empty, then the algorithm terminates with the exact value

$$R = R(f;X) \tag{5.1}$$

of the range of $f$ on $X$.

(b) If $r \geq N$ or $w(X_1) \leq \varepsilon$, then the algorithm terminates with the overestimate

$$R := R ++ F(X_1) ++ \ldots ++ F(X_r) \supset R(f;X) \tag{5.2}$$

of the range of f on X.

(c) Otherwise, each subinterval in $L_r$ is bisected to form a new list $L_n$ with n = 2r, and the algorithm returns to step 2°.

6. Remarks. The algorithm given in the previous section will terminate in a finite number of steps with either the exact value of R(f;X) or an overestimate which is never worse than

$$R = F(X_1) ++ \ldots ++ F(X_n) \supset R(f;X). \tag{6.1}$$

In general, (6.1) is a better approximation to R(f;X) than F(X) because of the convergence of united extensions to the range of a continuous function [1].

As a byproduct of the calculation when an overestimate is produced, the intervals $X_1,\ldots,X_r$ which are retained at the final step may contain critical points of f, that is, points at which f'(x) = 0. This information may be useful in optimization problems. Furthermore, if the list of retained intervals is nonempty, then the value $R \supset R(f;X)$ returned by the algorithm is definitely known to be an overestimate, while if the list of retained algorithms is empty, then this value is exact (modulo outward rounding). Thus, the algorithm itself indicates the type of result (exact or an overestimate) it obtains. The knowledge that R is an overestimate and the list of retained intervals can be used to refine the calculation of R(f;X) further, if desired. Some idea of the quality of the overestimate can be obtained by comparing the value of R before calculating (5.2) with the final result.

7. Numerical results. Numerical results were computed for the following functions, using the Pascal-SC program given in the following section.

$$f_1(x) = x - x, \tag{7.1}$$

$$f_2(x) = x \cdot x, \tag{7.2}$$

$$f_3(x) = \frac{(x - 1)\cdot(x + 3)}{(x + 2)}, \tag{7.3}$$

$$f_4(x) = x/x. \tag{7.4}$$

a. For X = [a,b], the naive interval extension $F_1(X) = X - X$ of $f_1$ gives F1([a,b]) = [a - b, b - a], while the algorithm gives R = [0,0] = $R(f_1;X)$ for arbitrary X.

b. For symmetric intervals $X = [-s,s]$, the algorithm gives the exact value $R = [0,s^2] = R(f_2; [-s,s])$, while $F_2([-s,s]) = [-s,s]\cdot[-s,s] = [-s^2,s^2]$. In case $X = [-r,s]$ is nonsymmetric interval containing 0, the result of the algorithm can be of the form $R = [-\varepsilon, \max\{r^2,s^2\}$, where $\varepsilon > 0$ is small, with a message that a small interval containing 0 can contain a critical point of $f_2$. For example, for $X = [-7,8]$, one has

$$F_2(X) = X\cdot X = [-56, 64] \tag{7.5}$$

while the algorithm gives

$$R = [-3.1\times10^{-18}, 64] \tag{7.6}$$

with a notation that there may be a critical point of $f_2$ in the retained interval $[-1.63\times10^{-9}, 1.87\times10^{-9}]$. In all other cases, the algorithm gives the exact result. Even if X is nonsymmetric about 0, the algorithm will give the correct result if 0 is a bisection point.

c. The function $f_3$ is actually monotone increasing, but has a pole at $x = -2$. The algorithm will sense the monotonicity of $f_3$ and give correct results if X is subdivided a sufficient number of times. The results are much better than the naive interval extension $F_3(X) = (X - 1)\cdot(X + 3)/(X + 2)$ when one of the endpoints of X is close to -2. For example, for $X = [-1.9, 98]$,

$$F_3(X) = [-2929, 97970], \tag{7.7}$$

while the algorithm gives

$$R = [-31.9, 97.97]. \tag{7.8}$$

For $X = [-1.999999, 98]$,

$$F_3(X) = [-3.03\times10^8, 9.797\times10^9], \tag{7.9}$$

while the algorithm gives

$$R = [-3000002, 97.97]. \tag{7.10}$$

Finally, for $X = [-1.99999999999, 98]$, which has a lower endpoint as close to -2 as possible in 12-digit decimal arithmetic, one gets

$$F_3(X) = [-3.03\times10^{13}, 9.797\times10^{14}], \tag{7.11}$$

while the algorithm gives

$$R = [-300000000002,\ 97.97] \tag{7.12}$$

d. The algorithm does not give good results for $f_4(x) = x/x$, because it determines that every subinterval of X possibly contains a critical point of $f_4$ (which in fact is true, since $f_4(x) \equiv 1$ is constant, and $f_4'(x) \equiv 0$). Thus, the algorithm computes R = [1,1] initially, and the final value is determined only by the united extension (6.1). Of course, the result is generally better than the naive interval extension $F_4(X) = X/X$ evaluated on the entire interval X, but is still usually a gross overestimate. For example, for X = [0.002, 2].

$$F_4(X) = [0.001,\ 1000] \tag{7.13}$$

while the algorithm gives

$$R = [0.203,\ 4.903], \tag{7.14}$$

which is still not a very good approximation to [1,1], even though it is much better than (7.13). Of course, the user is warned that the result may not be good by the fact that all subintervals are retained. Other methods usually give no warning when gross overestimates are produced. One way to improve the algorithm in this case, since interval extensions of derivatives are available, would be to use mean-value forms

$$F(X_i) = m(X_i) + F'(X_i)\cdot(X - m(X_i)) \tag{7.15}$$

to obtain interval extensions F of f on subintervals $X_i$, instead of obtaining them by straightforward evaluation.

8. A Pascal-SC program. The program written below was designed to be general, so that the user needs to supply only subroutines for evaluation of the function f in ordinary interval arithmetic (IFEVAL) and in interval differentiation arithmetic (IDFEVAL). The source code for these subroutines should be located in the files FEVAL.FUN. Examples of these subroutines for the functions discussed in §7 are given in §10.

The operators for interval differentiation arithmetic given in §9 include only the basic arithmetic operators for type IDERIV. For a complete computational system, operators for mixed arithmetic between types INTEGER, REAL, and IDERIV should be included, as well as standard functions [3].

The number of subintervals allowed in a list is set by the constant DIM in the program, which can be changed by the user. The size of the smallest subintervals is similarly controlled by the number LIMIT of bisections allowed. Thus, if LIMIT = L, then

$$\delta = 2^{-L} \cdot w(X), \tag{8.1}$$

where w(X) = b - a is the width of the original interval X = [a,b]. The source code for the Pascal-SC program follows:

```
PROGRAM IRANGE(INPUT,OUTPUT);
 CONST DIM = 256;        (* Maximum number of subintervals *)
       LIMIT = 32;       (* Maximum number of bisections *)
 TYPE INTERVAL = RECORD INF,SUP : REAL END ;
      IDERIV = RECORD X,PRIME: INTERVAL END;
      DIMTYPE = 1..DIM;
      STACKTYPE = RECORD INT:INTERVAL;FUN:IDERIV END;
 VAR X,RF,BEST,WORST: INTERVAL;
    F: IDERIV;
    I,NA,NB,LIM: INTEGER;
    A,B: ARRAY[DIMTYPE]OF STACKTYPE;  (* A is the list of intervals to
                                         be examined, B is the list of
                                         retained intervals *)
    MX: REAL;
$INCLUDE INTERVAL.PAK;          (* Makes interval arithmetic available *)
$INCLUDE IDERV.PAK;             (* Interval differentiation arithmetic *)
PROCEDURE IOUT(X: INTERVAL); (* Prints endpoints in standard format *)
 BEGIN
  WRITE('[',X.INF,',',X.SUP,']');
 END;
 $INCLUDE FEVAL.FUN;     (* Evaluation of the function in interval and
                            and interval differentiation arithmetic *)
FUNCTION RMF(L,G: REAL): INTERVAL;
 (* Bounds the range of a monotone function which assumes its least
    value at L and its greatest value at G. *)
 VAR D,U: INTERVAL;
 BEGIN
  D:=INTPT(L);U:=INTPT(G);
  D:=IFEVAL(D);U:=IFEVAL(U);
  D.SUP:=U.SUP;
  RMF:=D
 END;
```

```
FUNCTION MID(X: INTERVAL): REAL;  (* Calculates midpoint of an interval *)
 VAR A,B: ARRAY[1..2]OF REAL;
 BEGIN
  A[1]:=X.INF;B[1]:=0.5;
  A[2]:=X.SUP;B[2]:=0.5;
  MID:=SCALP(A,B,0)
 END;

BEGIN  (* Program IRANGE *)
 WRITELN('Enter initial interval X:');
 IREAD(INPUT,X);
 WRITE('    X = ');IOUT(X);WRITELN;
 F:=IDFEVAL(X);
 WORST:=F.X;
 IF (F.PRIME.INF >= 0) THEN RF:=RMF(X.INF,X.SUP)
 ELSE IF (F.PRIME.SUP <= 0) THEN RF:=RMF(X.SUP,X.INF)
 ELSE
 BEGIN  (* F is not monotone *)
  NA:=1;LIM:=0;
  A[1].INT:=X;A[1].FUN:=F;
  MX:=MID(X);
  X:=INTPT(MX);
  BEST:=IFEVAL(X);
  WHILE ((NA > 0) AND (NA <= DIM DIV 2) AND (LIM < LIMIT)) DO
  BEGIN (* WHILE *)
   LIM:=LIM+1;NB:=0;
   FOR I:=1 TO NA DO
    BEGIN (* STACK B *)
     MX:=MID(A[I].INT);
     NB:=NB+1;
     B[NB].INT.INF:=A[I].INT.INF;
     B[NB].INT.SUP:=MX;
     B[NB].FUN:=IDFEVAL(B[NB].INT);
     NB:=NB+1;
     B[NB].INT.INF:=MX;
     B[NB].INT.SUP:=A[I].INT.SUP;
     B[NB].FUN:=IDFEVAL(B[NB].INT);
    END;  (* STACK B *)
    NA:=0;
    FOR I:=1 TO NB DO
     BEGIN (* UNSTACK B *)
```

```
        IF NOT (B[I].FUN.X <= BEST)
        THEN IF (B[I].FUN.PRIME.INF >= 0)
        THEN BEST:=BEST+*RMF(B[I].INT.INF,B[I].INT.SUP)
        ELSE IF (B[I].FUN.PRIME.SUP <= 0)
        THEN BEST:=BEST+*RMF(B[I].INT.SUP,B[I].INT.INF)
        ELSE
         BEGIN (* RESTACK A *)
          NA:=NA+1;A[NA]:=B[I]
         END;  (* RESTACK A *)
       END;  (* UNSTACK B *)
       RF:=BEST;
       FOR I:=1 TO NA DO RF:=RF+*A[I].FUN.X;
    END;  (* WHILE *)
    IF NA > 0 THEN
    BEGIN (* NA > 0 *)
     RF:=BEST;
     WRITELN('Function may have critical points in:');
     FOR I:=1 TO NA DO
     BEGIN
      WRITE('A[',I:2,'] = ');IOUT(A[I].INT);WRITELN;
      RF:=RF+*A[I].FUN.X
     END;
    END;  (* NA > 0 *)
   END;   (* F is not monotone *)
   WRITELN('Naive interval arithmetic gives:');
   WRITE(' F(X) = ');IOUT(WORST);WRITELN;
   WRITELN('The algorithm gives:');
   WRITE(' F(X) = ');IOUT(RF);WRITELN
  END.  (* Program IRANGE *)
```

9. <u>The operators for interval differentiation arithmetic.</u> The six basic unary and binary arithmetic operators for type IDERIV are located in the file IDERIV.PAK, which also includes the call to the interval library for the function ISCALP to compute the interval scalar product.

```
TYPE IVECTOR = ARRAY[1..2]OF INTERVAL;
FUNCTION ISCALP ( VAR A,B: IVECTOR; DIM: INTEGER): INTERVAL;
 EXTERNAL 88;      (* Interval scalar product *)
OPERATOR + (U: IDERIV) RES: IDERIV;
 BEGIN
  RES:=U
 END;
```

```
OPERATOR - (U: IDERIV) RES: IDERIV;
 BEGIN
  U.X:=-U.X;
  U.PRIME:=-U.PRIME;
  RES:=U
 END;
OPERATOR + (U,V: IDERIV) RES: IDERIV;
 BEGIN
  U.X:=U.X+V.X;
  U.PRIME:=U.PRIME+V.PRIME;
  RES:=U
 END;
OPERATOR - (U,V: IDERIV) RES: IDERIV;
 BEGIN
  U.X:=U.X-V.X;
  U.PRIME:=U.PRIME-V.PRIME;
  RES:=U
 END;
OPERATOR * (U,V: IDERIV) RES: IDERIV;
 VAR A,B: IVECTOR;
 BEGIN
  A[1]:=U.X;B[1]:=V.PRIME;
  A[2]:=V.X;B[2]:=V.PRIME;
  U.PRIME:=ISCALP(A,B,2);
  U.X:=U.X*V.X;
  RES:=U
 END;

OPERATOR / (U,V: IDERIV) RES: IDERIV;
 VAR A,B: IVECTOR;
     C: IDERIV;
 BEGIN
  C.X:=U.X/V.X;
  A[1]:=INTPT(1);B[1]:=U.PRIME;
  A[2]:=-C.X;B[2]:=V.PRIME;
  C.PRIME:=ISCALP(A,B,2)/V.X;
  RES:=C
END;
```

10. Example function subroutines. (Contents of the file FEVAL.FUN.)

(a) $f_1(x) = x - x$.

```
FUNCTION IFEVAL(X: INTERVAL): INTERVAL;
 BEGIN
  IFEVAL := X - X
 END;
FUNCTION IDFEVAL(X: IDERIV): IDERIV;
 BEGIN
  IDFEVAL := X - X
 END;
```

(b) $f_2(x) = x \cdot x$.

```
FUNCTION IFEVAL(X: INTERVAL): INTERVAL;
 BEGIN
  IFEVAL := X*X
 END;
FUNCTION IDFEVAL(X: IDERIV): IDERIV;
  BEGIN
   IDFEVAL := X*X
  END
```

(c) $f_3(x) = (x - 1)\cdot(x + 3)/(x + 2)$.

```
FUNCTION IFEVAL(X: INTERVAL): INTERVAL;
 BEGIN
  IFEVAL := (X - 1)*(X + 3)/(X + 2)
 END;
FUNCTION IDFEVAL(X: IDERIV): IDERIV;
 VAR ONE,TWO,THREE: IDERIV;
 BEGIN
  ONE.X := INTPT(1);ONE.PRIME := INTPT(0);
  TWO.X := INTPT(2);TWO.PRIME := INTPT(0);
  THREE.X := INTPT(3);THREE.PRIM := INTPT(0);
  IDFEVAL := (X - ONE)*(X + THREE)/(X + TWO)
 END;
```

(d) $f_4(x) = x/x$.

```
FUNCTION IFEVAL(X: INTERVAL): INTERVAL;
 BEGIN
  IFEVAL := X/X
 END;
FUNCTION IDFEVAL(X: IDERIV): IDERIV;
 BEGIN
```

```
IDFEVAL := X/X
END;
```

11. Acknowledgements. This work was sponsored by the United States Army under Contract No. DAAG29-80-C-0041. The author is grateful to Prof. Karl Nickel for helpful comments and suggestions.

References

1. R. E. Moore. Interval Analysis, Prentice-Hall, Englewood Cliffs, N. J., 1966.

2. L. B. Rall. Automatic Differentiation: Techniques and Applications. Lecture Notes in Computer Science No. 120, Springer, New York, 1981.

3. L. B. Rall. Differentiation and generation of Taylor coefficients in Pascal-SC, pp. 291-309 in A New Approach to Scientific Computation, ed. by U. W. Kulisch and W. L. Miranker, Academic Press, New York, 1983.

4. L. B. Rall. The Arithmetic of Differentiation. MRC Technical Summary Report No. 2688, Mathematics Research Center, University of Wisconsin-Madison, May, 1984. To appear in Mathematics Magazine, February, 1986.

5. H. Ratschek and J. Rokne. Computer Methods for the Ranges of Functions. Halsted Press, Wiley, New York, 1984.

## INNER SOLUTIONS OF LINEAR INTERVAL SYSTEMS

J. Rohn
Charles University
Prague/Czechoslovakia

A vector $x \in R^n$ is called an inner solution of a system of linear interval equations $A^I x = b^I$ ($A^I = [\underline{A},\overline{A}] = [A_c - \Delta, A_c + \Delta]$ of size $m \times n$, $b^I = [\underline{b},\overline{b}] = [b_c - \delta, b_c + \delta]$) if $Ax \in b^I$ for each $A \in A^I$ (for a motivation, see [1]). Denote by $X_i$ the set of all inner solutions. We have this characterization:

Theorem. $x \in X_i$ if and only if $x = x_1 - x_2$, where $x_1, x_2$ is a solution to the system of linear inequalities

$$\begin{aligned} \overline{A}x_1 - \underline{A}x_2 &\leqq \overline{b} \\ -\underline{A}x_1 + \overline{A}x_2 &\leqq -\underline{b} \\ x_1 \geqq 0, \quad x_2 &\geqq 0. \end{aligned} \tag{S}$$

Proof. Due to Oettli-Prager theorem, $\{Ax;\ A \in A^I\} = [A_c x - \Delta|x|, A_c x + \Delta|x|]$. "Only if": Let $x \in X_i$, then $\underline{b} \leqq A_c x - \Delta|x|$ and $A_c x + \Delta|x| \leqq \overline{b}$; substituting $x = x^+ - x^-$, $|x| = x^+ - x^-$, we see that $x_1 = x^+$, $x_2 = x^-$ satisfy (S). "If": Let $x_1, x_2$ solve (S); define $d \in R^n$ by $d_j = \min\{x_{1j}, x_{2j}\}$ $\forall$ j, then $d \geqq 0$ for $x = x_1 - x_2$ we have $x^+ = x_1 - d$, $x^- = x_2 - d$, hence $A_c x + \Delta|x| = \overline{A}x_1 - \underline{A}x_2 - 2\Delta d \leqq \overline{b}$, similarly $A_c x - \Delta|x| \geqq \underline{b}$. Thus $[A_c x - \Delta|x|, A_c x + \Delta|x|] \subset b^I$, implying $x \in X_i$. ⊠

As consequences, we obtain: (i) $X_i$ is a convex polytope, (ii) each $x \in X_i$ satisfies $\Delta|x| \leqq \delta$ (by adding the first two inequalities in (S)), (iii) $X_i$ is bounded if for each j there is a k with $\Delta_{kj} > 0$ (since then from (ii) follows $|x_j| \leqq \delta_k/\Delta_{kj}$), (iv) $X_i \neq \emptyset$ if and only if (S) has a solution, which can be tested by phase I of the simplex algorithm, (v) for $\underline{x}_j = \min\{x_j;\ x \in X_i\}$ we have $\underline{x}_j = \min\{(x_1 - x_2)_j;\ x_1, x_2 \text{ solve (S)}\}$, which is a linear programming problem (similarly for $\overline{x}_j = \max\ldots$), (vi) nonnegative inner solutions are described by $\overline{A}x \leq \overline{b}$, $-\underline{A}x \leq -\underline{b}$, $x \geq 0$, (vii) also, $X_i = \{x;\ |A_c x - b_c| \leq -\Delta|x| + \delta\}$ (observe the similarity with the Oettli-Prager result).

Acknowledgment.

Dr. A. Deif's posing the problem to the author is acknowledged with thanks.

Reference.

[1] Nuding, E.; Wilhelm, J.: Über Gleichungen und über Lösungen, ZAMM 52, T188 - T190 (1972).

# EMBEDDING THEOREMS FOR CONES AND APPLICATIONS TO CLASSES OF CONVEX SETS OCCURRING IN INTERVAL MATHEMATICS

Klaus D. Schmidt

Seminar für Statistik, Universität Mannheim, A 5
6800 Mannheim, West Germany

ABSTRACT

This paper gives a survey of embedding theorems for cones and their application to classes of convex sets occurring in interval mathematics.

## 1. INTRODUCTION

In many situations, the investigation of set-valued maps can be reduced to the vector-valued case by applying embedding theorems for classes of convex sets. For example, Rådström's embedding theorem for the class of all nonempty, compact, convex subsets of a normed vector space has been used in the construction of the Debreu integral and in the proof of a law of large numbers for random sets, and there is some hope that such embedding theorems can also be used for proving fixed-point theorems for set-valued maps which are needed in interval mathematics and other areas like mathematical economics.

An interesting method for proving embedding theorems for classes of convex sets is that of Rådström [10] who first established the cone properties of the class of convex sets under consideration and then applied a general embedding theorem for cones to prove his embedding theorems for the class of all nonempty, compact, convex subsets of a normed vector space and for the class of all nonempty, closed, bounded, convex subsets of a reflexive Banach space. Rådström's method has also been used by Urbanski [15] who considered a more general situation, and it has quite recently been used by Fischer [1] who proved an embedding theorem for the class of all hypernorm balls of a hypernormed vector space which can be applied to norm balls and order intervals.

Since the value of such embedding theorems for classes of convex sets depends on the amount of information they provide on the embedding vector space and the embedding map, it is desirable to have embedding theorems which also reflect the inclusion of sets as an order relation as well as the relationship existing between the topological and order properties of the class of convex sets under consideration. For example, taking into account the inclusion of sets as an order relation has led to more informative versions of the embedding theorems proven by Rådström [10], Hörmander [2], and Fischer [1], and it has also led to new embedding theorems for the class of all order intervals of an (M-normed) vector lattice (with unit); see [13] and [14]. With regard to Rådström's method for proving embedding theorems for classes of convex sets, the results of [13] and [14] suggest a systematic study of embedding theorems for (topological) ordered cones.

The purpose of these notes is to give a survey of embedding theorems for cones and their application to classes of convex sets occuring in interval mathematics.

In Section 2, we present some known and several new embedding theorems fo cones, ordered cones, topological cones, and topological ordered cones. As far as topological properties are concerned, we confine ourselves to the case where the topology is determined by a positively homogeneous translation-invariant metric. For applications in interval mathematics, this seems to be a reasonable restriction, as remarked by Ratschek [11].

In Section 3, we study the cone properties of norm balls, hypernorm balls and order intervals. These classes of convex sets are always endowed with the Minkowski addition of sets, the usual multiplication of sets by positive scalars, and the inclusion of sets as an order relation, and the metrics under consideration are those of Hausdorff and Moore. Using the results of Section 3 and applying suitable embedding theorems for cones given in Section 2, it is not hard to establish embedding theorems for norm balls, hypernorm balls, and order intervals. For the brevity of the presentation in these notes, however, the formulation of the resulting embedding theorems must be left to the reader, but we remark that some of them may be found in [14]; see also [13] for intervals on the real line.

In Section 4, we indicate some further aspects of embedding theorems for classes of convex sets occuring in interval mathematics. In particular, we briefly discuss the relationship between quasilinear spaces and cones we sketch the cone properties of order intervals with respect to an orde relation which differs from the inclusion of sets but has the advantage of extending the order relation of the underlying vector lattice, and we also include some comments on concrete embedding theorems for order intervals.

For any details concerning ordered vector spaces and (normed) vector lattices, we refer to the books by Luxemburg and Zaanen [6] and by Schaefer [12].

## 2. EMBEDDING THEOREMS FOR CONES

In this section, we present some known and several new embedding theorems for cones, ordered cones, normed cones, and normed ordered cones. For the formulation of these embedding theorems, we have to introduce some new terminology for cones. Although some of the new definitions we introduce may be tentative, they are convenient for our purposes, and all of them are in accordance with the corresponding definitions for vector spaces. The proofs of the embedding theorems for cones are somewhat technical and lengthy and cannot be included in these notes. However, to give an idea of the proofs, we indicate the construction of the embedding vector space and the embedding map. Some further details may be found in the papers of Rådström [10] and Kaucher [4]; see also [13] and [14].

### C o n e s

A cone (or semilinear space [1]) is a set $\mathbb{F}$ with a distinguished element $Z \in \mathbb{F}$ (called the zero element), a map $+ : \mathbb{F}\times\mathbb{F} \longrightarrow \mathbb{F}$ (called addition) satisfying $A+(B+C) = (A+B)+C$ , $A+B = B+A$ , and $A+Z = A$ for all $A, B, C \in \mathbb{F}$ , and a map $\mathbb{R}_+\times\mathbb{F} \longrightarrow \mathbb{F}$ (called scalar multiplication) satisfying $\lambda(A+B) = \lambda A+\lambda B$ , $(\lambda+\mu)A = \lambda A+\mu A$ , $(\lambda\mu)A = \lambda(\mu A)$ , $1A = A$ , and $0A = Z$ for all $A, B \in \mathbb{F}$ and $\lambda, \mu \in \mathbb{R}_+$ . If $\mathbb{F}$ is a cone with zero element $Z$ , then the identity $\lambda Z = Z$ holds for all $\lambda \in \mathbb{R}_+$ . A cone $\mathbb{F}$ has the cancellation property if $A = B$ holds for all $A, B \in \mathbb{F}$ satisfying $A+C = B+C$ for some $C \in \mathbb{F}$ .

Let $\mathbb{F}$ be a cone having the cancellation property.
On $\mathbb{F}\times\mathbb{F}$ , define an equivalence relation $\sim$ by letting

$$(A,B) \sim (C,D)$$

for all $(A,B), (C,D) \in \mathbb{F}\times\mathbb{F}$ , and let

$$\langle A,B \rangle$$

denote the equivalence class containing $(A,B)$ .
Let $\mathbb{G}$ denote the collection of all equivalence classes of $\mathbb{F}\times\mathbb{F}$ .

On $\mathbb{G}$ , define addition $+ : \mathbb{G}\times\mathbb{G} \longrightarrow \mathbb{G}$ and scalar multiplication $\mathbb{R}\times\mathbb{G} \longrightarrow \mathbb{G}$ by letting

$$\langle A,B\rangle + \langle C,D\rangle := \langle A+C,B+D\rangle$$

and

$$\alpha\langle A,B\rangle := \begin{cases} \langle \alpha A , \alpha B \rangle , & \text{if } \alpha \in \mathbb{R}_+ \\ \langle(-\alpha)B,(-\alpha)A\rangle , & \text{otherwise} \end{cases}$$

for all $\langle A,B\rangle, \langle C,D\rangle \in \mathbb{G}$ and $\alpha \in \mathbb{R}$ .

Furthermore, define a map $j : \mathbb{F} \longrightarrow \mathbb{G}$ by letting

$$j(A) := \langle A,Z\rangle$$

for all $A \in \mathbb{F}$ .

Then we have the following basic embedding theorem for cones, which is due to Rådström [10]:

**2.1. Theorem.**

Suppose $\mathbb{F}$ is a cone having the cancellation property. Then $\mathbb{G}$ is a vector space satisfying $\mathbb{G} = j(\mathbb{F}) - j(\mathbb{F})$ , and $j$ is an injection which is additive and positively homogeneous. In particular, the identities $\langle A,B\rangle + \langle Z,Z\rangle = \langle A,B\rangle$ and $\langle A,B\rangle + \langle B,A\rangle = \langle Z,Z\rangle$ hold for all $\langle A,B\rangle \in \mathbb{G}$ .

In the formulation of subsequent embedding theorems, we shall usually not repeat those properties of the embedding vector space $\mathbb{G}$ and the embedding map $j$ which are evident from more general results.

## Ordered Cones

An ordered cone is a cone $\mathbb{F}$ with an order relation $\le$ such that $A + C \le B + C$ and $\lambda A \le \lambda B$ holds for all $A, B \in \mathbb{F}$ satisfying $A \le B$ and for all $C \in \mathbb{F}$ and $\lambda \in \mathbb{R}_+$ . If $\mathbb{F}$ is an ordered cone, then the set $\mathbb{F}_+ := \{ A \in \mathbb{F} \mid Z \le A \}$ is said to be the positive cone of $\mathbb{F}$ . An ordered cone $\mathbb{F}$ has the order cancellation property if $A \le B$ holds for all $A, B \in \mathbb{F}$ satisfying $A + C \le B + C$ for some $C \in \mathbb{F}$ , it is Archimedean if $A \le B$ holds for all $A, B \in \mathbb{F}$ satisfying $nA + D \le nB + C$ for some $C, D \in \mathbb{F}$ and all $n \in \mathbb{N}$ , and it has the Hukuhara property if for all $A, B \in \mathbb{F}$ satisfying $A \le B$ there exists some $D \in \mathbb{F}_+$ satisfying $A + D = B$ .

Each Archimedean ordered cone has the order cancellation property, and each ordered cone having the cancellation property and the Hukuhara property has the order cancellation property.

Let $\mathbb{F}$ be an ordered cone having the order cancellation property.

On $\mathbb{G}$ , define an order relation $\leq$ by letting

$$\langle A,B\rangle \leq \langle C,D\rangle$$

for all $\langle A,B\rangle, \langle C,D\rangle \in \mathbb{G}$ satisfying $A + D \leq B + C$ .

2.2. Theorem.

Suppose $\mathbb{F}$ is an ordered cone having the order cancellation property. Then $\mathbb{G}$ is an ordered vector space, and $j$ is isotone and inverse-isotone.
Moreover, $\mathbb{G}$ is Archimedean if and only if $\mathbb{F}$ is Archimedean, and $\mathbb{G}_+ = j(\mathbb{F}_+)$ holds if and only if $\mathbb{F}$ has the Hukuhara property.

A semilattice cone (or upper semilattice cone [13,14]) is an ordered cone $\mathbb{F}$ such that $A \vee B := \sup\{A,B\}$ exists for all $A, B \in \mathbb{F}$ and the identity $(A+C)\vee(B+C) = A\vee B + C$ holds for all $A, B, C \in \mathbb{F}$ .
Each semilattice cone having the cancellation property has the order cancellation property, a semilattice cone $\mathbb{F}$ is Archimedean if and only if $A = B$ holds for all $A, B \in \mathbb{F}$ satisfying $B \leq A$ and $nA + D \leq nB + C$ for some $C, D \in \mathbb{F}$ and all $n \in \mathbb{N}$ , and a semilattice cone $\mathbb{F}$ having the cancellation property and the Hukuhara property is Archimedean if and only if $A = Z$ holds for all $A \in \mathbb{F}_+$ satisfying $nA \leq C$ for some $C \in \mathbb{F}$ and all $n \in \mathbb{N}$ .

A semilattice cone $\mathbb{F}$ has the Riesz property if the identity $A + B = A\vee B + A\wedge B$ holds for all $A, B \in \mathbb{F}$ for which $A\wedge B := \inf\{A,B\}$ exists. If $\mathbb{F}$ is a semilattice cone having the cancellation property and the Riesz property, then the identity $(A+C)\wedge(B+C) = A\wedge B + C$ holds for all $A, B, C \in \mathbb{F}$ for which $(A+C)\wedge(B+C)$ and $A\wedge B$ exist.

2.3. Theorem.

Suppose $\mathbb{F}$ is a semilattice cone having the cancellation property. Then $\mathbb{G}$ is a vector lattice, and $j$ preserves finite suprema.
In particular, the identities $\langle A,B\rangle \vee \langle C,D\rangle = \langle (A+D)\vee(B+C), B+D\rangle$ and $\langle A,B\rangle \wedge \langle C,D\rangle = \langle A+C, (A+D)\vee(B+C)\rangle$ hold for all $\langle A,B\rangle, \langle C,D\rangle \in \mathbb{G}$ .
Moreover, $j$ preserves finite infima if and only if $\mathbb{F}$ has the Riesz property.

An ordered cone $\mathbb{F}$ is (countably) order complete if $\sup\{A_\gamma \mid \gamma \in \Gamma\}$ exists for each (countable) set $\{A_\gamma \in \mathbb{F} \mid \gamma \in \Gamma\}$ satisfying $A_\gamma \leq C$ for some $C \in \mathbb{F}$ and all $\gamma \in \Gamma$ .
Each countably order complete semilattice cone having the cancellation property and the Hukuhara property is Archimedean.

2.4. Theorem.

Suppose $\mathbb{F}$ is a semilattice cone having the cancellation property and the Hukuhara property.
Then $\mathbb{G}$ is (countably) order complete if and only if $\mathbb{F}$ is (countably) order complete.

In the case where $\mathbb{F}$ is only an ordered semigroup, some of the previous results have been proven by Kaucher [4].

## Normed Cones

If $\mathbb{F}$ is a cone and $d : \mathbb{F}\times\mathbb{F} \longrightarrow \mathbb{R}_+$ is a positively homogeneous translation-invariant metric, then

(i) $d(A,Z) = 0$ if and only if $A = Z$,

(ii) $d(A+B,Z) \leq d(A,Z) + d(B,Z)$, and

(iii) $d(\lambda A,Z) = \lambda d(A,Z)$

holds for all $A, B \in \mathbb{F}$ and $\lambda \in \mathbb{R}_+$, and

(iv) $d(A,Z) = d(B,Z)$

holds for all $A, B \in \mathbb{F}$ satisfying $A + B = Z$. These properties of the map $d(.,Z) : \mathbb{F} \longrightarrow \mathbb{R}_+$ suggest the following definition of a normed cone, but it should be noted that the existence of a map $\mathbb{F} \longrightarrow \mathbb{R}_+$ having properties (i) - (iv) is not equivalent to the existence of a positively homogeneous translation-invariant metric on $\mathbb{F}$; see also Mayer [7].

A normed cone is a cone $\mathbb{F}$ with a metric $d$ which is translation-invariant and positively homogeneous. If $\mathbb{F}$ is a normed cone, then addition and scalar multiplication on $\mathbb{F}$ are continuous.
Each normed cone has the cancellation property.

Let $\mathbb{F}$ be a normed cone.
On $\mathbb{G}$, define a norm $\|.\| : \mathbb{G} \longrightarrow \mathbb{R}_+$ by letting

$$\| \langle A,B\rangle \| := d(A,B)$$

for all $\langle A,B\rangle \in \mathbb{G}$.

2.5. Theorem.

Suppose $\mathbb{F}$ is a normed cone.
Then $\mathbb{G}$ is a normed vector space, and $j$ is isometric.

The previous result is due to Rådström [10]. For an extension of Theorem 2.5 to more general topological cones, see Urbanski [15].

## Normed Ordered Cones

A normed ordered cone is an ordered cone $\mathbb{F}$ with a metric $d$ which is translation-invariant and positively homogeneous and satisfies one (and thus all) of the following equivalent conditions:

- $d(A,B) \leq d(A,C)$ holds for all $A, B, C \in \mathbb{F}$ satisfying $A \leq B \leq C$ ;
- $d(B,C) \leq d(A,C)$ holds for all $A, B, C \in \mathbb{F}$ satisfying $A \leq B \leq C$ ;
- $d(A,B) \leq d(C,D)$ holds for all $A, B, C, D \in \mathbb{F}$ satisfying $B \leq A$ , $D \leq C$ , and $A + D \leq B + C$ .

A metric satisfying these conditions is said to be chain isotone.

2.6. Theorem.
Suppose $\mathbb{F}$ is a normed ordered cone having the order cancellation property.
Then $\mathbb{G}$ is a normed ordered vector space.

A normed semilattice cone is a semilattice cone $\mathbb{F}$ with a metric $d$ which is translation-invariant and positively homogeneous and satisfies one (and thus all) of the following equivalent conditions:

- $d$ is chain isotone and $d((A+A)\vee(B+B), A+B) = d(A,B)$ holds for all $A, B \in \mathbb{F}$ ;
- $d(A,B) \leq d(C,D)$ holds for all $A, B, C, D \in \mathbb{F}$ satisfying $(A+A)\vee(B+B) + C + D \leq A + B + (C+C)\vee(D+D)$ .

Each normed semilattice cone is Archimedean; in particular, each normed semilattice cone has the order cancellation property.

An M-normed semilattice cone (with unit) is a normed semilattice cone $\mathbb{F}$ such that

- $d((A+D)\vee(B+C), B+D) = \max\{ d(A,B) , d(C,D) \}$ holds for all $A, B, C, D \in \mathbb{F}$ satisfying $B \leq A$ and $D \leq C$ (and there exists some $E \in \mathbb{F}$ satisfying $d(E,Z) = 1$ and $A \leq B + E$ for all $A, B \in \mathbb{F}$ satisfying $d(A,B) \leq 1$ ).

2.7. Theorem.
Suppose $\mathbb{F}$ is a normed semilattice cone.
Then $\mathbb{G}$ is a normed vector lattice.
Moreover, $\mathbb{G}$ is an M-normed vector lattice (with unit) if and only if $\mathbb{F}$ is an M-normed semilattice cone (with unit).

The previous results can be extended to more general topological ordered cones, as indicated in [13].

## 3. CONES OF CONVEX SETS OCCURING IN INTERVAL MATHEMATICS

In this section, we study the cone properties of the classes of all norm balls of a normed vector space, all hypernorm balls of a hypernormed vector space, and all order intervals of a (normed) vector lattice. The cone properties of these classes of convex sets are studied with respect to the Minkowski addition of sets $+$ which is defined by letting

$$A + B := \{ a + b \mid a \in A , b \in B \}$$

for all nonempty subsets $A$ and $B$ of a (real) vector space $\mathbb{E}$, the usual multiplication of sets by positive scalars which is defined by letting

$$\lambda A := \{ \lambda a \mid a \in A \}$$

for all nonempty subsets $A$ of $\mathbb{E}$ and (positive) scalars $\lambda \in \mathbb{R}_+$, and the inclusion of sets $\subseteq$. The cone properties of the class of all norm balls of a normed vector space $\mathbb{E}$ are also studied with respect to the Hausdorff distance $\Delta$ which is defined by letting

$$\Delta(A,B) := \max \{ \sup_A \inf_B \|a-b\| , \sup_B \inf_A \|b-a\| \}$$

for all nonempty bounded subsets $A$ and $B$ of $\mathbb{E}$, and those of the class of all order intervals of a normed vector lattice $\mathbb{E}$ are also studied with respect to the Moore distance $\delta$ which is defined by letting

$$\delta([a,b],[c,d]) := \max \{ \|a-c\| , \|b-d\| \}$$

for all order intervals $[a,b]$ and $[c,d]$ of $\mathbb{E}$ and which in certain cases agrees with the Hausdorff distance $\Delta$. The proofs of the cone properties of these classes of convex sets cannot be included in these notes, but parts of them may be found in the paper of Fischer [1] and in [13] and [14].

### Norm Balls

If $\mathbb{E}$ is a normed vector space with norm $\|.\|$, let

$$\overline{U} := \{ x \in \mathbb{E} \mid \|x\| \leq 1 \}$$

denote the closed unit ball of $\mathbb{E}$. It is known and can be proven by a slight modification of the proof of [13; Theorem 6.2] that the identity

$$\Delta(A,B) = \inf \{ \varepsilon \in (0,\infty) \mid A \subseteq B + \varepsilon\overline{U} , B \subseteq A + \varepsilon\overline{U} \}$$

holds for all nonempty bounded subsets $A$ and $B$ of $\mathbb{E}$.

Let $\mathbb{E}$ be a normed vector space with norm $\|.\|$. A subset $A$ of $\mathbb{E}$ is a norm ball of $\mathbb{E}$ if there exist $m \in \mathbb{E}$ and $\mu \in \mathbb{R}_+$ satisfying

$$\langle m;\mu\rangle := \{ x \in \mathbb{E} \mid \|x-m\| \leq \mu \} = m + \mu\overline{U} = A .$$

This midpoint-radius representation of a norm ball is clearly unique. Let $\mathbb{F}_b(\mathbb{E}, \|.\|)$ denote the class of all norm balls of $\mathbb{E}$, endowed with the Minkowski addition of sets, the usual multiplication of sets by positive scalars, and the inclusion of sets.

3.1. Theorem.

Suppose $\mathbb{E}$ is a normed vector space.
Then $\mathbb{F}_b(\mathbb{E}, \|.\|)$ is an Archimedean ordered cone having the Hukuhara property.
In particular, the identity $\langle m_1;\mu_1\rangle + \langle m_2;\mu_2\rangle = \langle m_1+m_2;\mu_1+\mu_2\rangle$ holds for all norm balls $\langle m_1;\mu_1\rangle$ and $\langle m_2;\mu_2\rangle$ of $\mathbb{E}$ .

We remark, however, that $\mathbb{F}_b(\mathbb{E}, \|.\|)$ may fail to be a semilattice cone; see [14; Example 5.4].

The following lemma can be used to give a simple proof of the properties of the Hausdorff distance on the class of all norm balls of $\mathbb{E}$ :

3.2. Lemma.

Suppose $\mathbb{E}$ is a normed vector space.
Then the identity

$$\Delta(\langle m_1;\mu_1\rangle , \langle m_2;\mu_2\rangle) = \| m_1 - m_2 \| + | \mu_1 - \mu_2 |$$

holds for all norm balls $\langle m_1;\mu_1\rangle$ and $\langle m_2;\mu_2\rangle$ of $\mathbb{E}$ .

Let $\mathbb{F}_b^\Delta(\mathbb{E}, \|.\|)$ denote the ordered cone $\mathbb{F}_b(\mathbb{E}, \|.\|)$ endowed with the Hausdorff distance $\Delta$ .

3.3. Theorem.

Suppose $\mathbb{E}$ is a normed vector space.
Then $\mathbb{F}_b^\Delta(\mathbb{E}, \|.\|)$ is an Archimedean normed ordered cone having the Hukuhara property.

This result can be proven by using Lemma 3.2 or [13; Lemma 3.2].

## Hypernorm Balls

If $\mathbb{E}$ is a vector space and $\mathbb{P}$ is an ordered vector space, then a map $h : \mathbb{E} \longrightarrow \mathbb{P}_+$ is a $\mathbb{P}$-hypernorm on $\mathbb{E}$ or briefly a hypernorm [1] if

(i) $h(x) = 0$ if and only if $x = 0$ ,

(ii) $h(x+y) \leq h(x) + h(y)$ , and

(iii) $h(\alpha x) = |\alpha| h(x)$

holds for all $x, y \in \mathbb{E}$ and $\alpha \in \mathbb{R}_+$ .
A $\mathbb{P}$-hypernorm $h$ on a vector space $\mathbb{E}$ is splittable if for all $x \in \mathbb{E}$ satisfying $h(x) \leq p + q$ for some $p, q \in \mathbb{P}_+$ there exist $y, z \in \mathbb{E}$ satisfying $x = y + z$ , $h(y) \leq p$ , and $h(z) \leq q$ , and it is surjective

if for all $p \in \mathbb{P}_+$ there exists some $x \in \mathbb{E}$ satisfying $h(x) = p$ .

A <u>$\mathbb{P}$-hypernormed vector space</u> or briefly a <u>hypernormed vector space</u> is a vector space $\mathbb{E}$ with a $\mathbb{P}$-hypernorm $h$ and will be denoted by $(\mathbb{E}, h)$ . For example, if $\mathbb{E}$ is a normed vector space with norm $\|.\|$ , then $(\mathbb{E}, \|.\|)$ is an $\mathbb{R}$-hypernormed vector space, and if $\mathbb{E}$ is a vector lattice with modulus $|.|$ , then $(\mathbb{E}, |.|)$ is an $\mathbb{E}$-hypernormed vector space, and in either case the hypernorm is splittable and surjective.

Let $(\mathbb{E}, h)$ be a $\mathbb{P}$-hypernormed vector space. A subset $A$ of $\mathbb{E}$ is a <u>hypernorm ball</u> of $\mathbb{E}$ if there exist $m \in \mathbb{E}$ and $p \in \mathbb{P}_+$ satisfying

$$\langle m;p\rangle := \{ x \in \mathbb{E} \mid h(x-m) \leq p \} = A .$$

Different from the case of norm balls, however, this <u>midpoint-radius representation</u> of a hypernorm ball need not be unique, as pointed out by Fischer [1]. Let $\mathbb{F}_b(\mathbb{E}, h)$ denote the class of all hypernorm balls of $\mathbb{E}$, endowed with the Minkowski addition of sets, the usual multiplication of sets by positive scalars, and the inclusion of sets.

<u>3.4. Theorem.</u>

Suppose $(\mathbb{E}, h)$ is a $\mathbb{P}$-hypernormed vector space such that $\mathbb{P}$ is Archimedean and $h$ is splittable.

Then $\mathbb{F}_b(\mathbb{E}, h)$ is an Archimedean ordered cone.

In particular, the identity $\langle m_1;p_1\rangle + \langle m_2;p_2\rangle = \langle m_1+m_2;p_1+p_2\rangle$ holds for all hypernorm balls $\langle m_1;p_1\rangle$ and $\langle m_2;p_2\rangle$ of $(\mathbb{E}, h)$ .

The previous result improves [14; Theorem 5.3] where a weaker definition of an Archimedean ordered cone has been used; see also Fischer [1]. Under an additional assumption on the hypernorm, Theorem 3.4 can be improved as to yield the following complete extension of Theorem 3.1:

<u>3.5. Theorem.</u>

Suppose $(\mathbb{E}, h)$ is a $\mathbb{P}$-hypernormed vector space such that $\mathbb{P}$ is Archimedean and $h$ is splittable and surjective.

Then $\mathbb{F}_b(\mathbb{E}, h)$ is an Archimedean ordered cone having the Hukuhara property.

Since hypernormed vector spaces generalize normed vector spaces, $\mathbb{F}_b(\mathbb{E}, h$ may fail to be a semilattice cone, by the remark following Theorem 3.1.

It would be interesting to know whether Theorem 3.3 can be extended to th hypernormed case by replacing the norm in the definition of the Hausdorff distance by the hypernorm $h$ (in the case where $\mathbb{P}$ is order complete).

In [14], the corresponding question for a different class of convex sets in a hypernormed vector space has been answered in the negative.

## Order Intervals

Let $\mathbb{E}$ be a vector lattice with modulus $|.|$ . A subset $A$ of $\mathbb{E}$ is an order interval of $\mathbb{E}$ if there exist $a, b \in \mathbb{E}$ satisfying $a \le b$ and

$$[a,b] := \{ x \in \mathbb{E} \mid a \le x \le b \} = A .$$

This lower-bound-upper-bound representation of an order interval is unique and yields a (unique) midpoint-radius representation of an order interval with respect to the modulus $|.|$ ; see e.g. [14; Proposition 6.1].
Let $\mathbb{F}_b(\mathbb{E}, |.|)$ denote the class of all order intervals of $\mathbb{E}$ , endowed with the Minkowski addition of sets, the usual multiplication of sets by positive scalars, and the inclusion of sets.

3.6. Theorem.
Suppose $\mathbb{E}$ is a vector lattice.
Then $\mathbb{F}_b(\mathbb{E}, |.|)$ is a semilattice cone having the cancellation property, the Hukuhara property, and the Riesz property.
In particular, the identities $[a,b] + [c,d] = [a+c,b+d]$ and $[a,b] \vee [c,d] = [a\wedge c,b\vee d]$ hold for all order intervals $[a,b]$ and $[c,d]$ of $\mathbb{E}$ , and the identity $[a,b] \wedge [c,d] = [a\vee c,b\wedge d]$ holds for all order intervals $[a,b]$ and $[c,d]$ of $\mathbb{E}$ having nonempty intersection.
Moreover, $\mathbb{F}_b(\mathbb{E}, |.|)$ is Archimedean if and only if $\mathbb{E}$ is Archimedean, and $\mathbb{F}_b(\mathbb{E}, |.|)$ is (countably) order complete if and only if $\mathbb{E}$ is (countably) order complete.

The previous result has been proven in [14; Lemma 6.2, Lemma 6.3, Lemma 6.5, and Theorem 6.6]. In the case where $\mathbb{E}$ is Archimedean, some of the assertions of Theorem 3.6 can be obtained from Theorem 3.5, but even in that case Theorem 3.6 provides more information on $\mathbb{F}_b(\mathbb{E}, |.|)$ than Theorem 3.5 does. Therefore, it seems not to be convenient to consider the order intervals of $\mathbb{E}$ as hypernorm balls of $(\mathbb{E}, |.|)$ .

Let now $\mathbb{E}$ be a normed vector lattice with modulus $|.|$ and norm $\|.\|$ , and let $\mathbb{F}_b^\delta(\mathbb{E}, |.|)$ denote the ordered cone $\mathbb{F}_b(\mathbb{E}, |.|)$ endowed with the Moore distance $\delta$ . The Moore distance has been introduced by Moore [8] in the case $\mathbb{E} = \mathbb{R}$ , and it has been used by Jahn [3] in the general case.

3.7. Theorem.
Suppose $\mathbb{E}$ is a normed vector lattice.
Then $\mathbb{F}_b^\delta(\mathbb{E}, |.|)$ is a normed semilattice cone.

Moreover, $\mathbb{F}_b^{\delta}(\mathbb{E}, |.|)$ is an M-normed semilattice cone (with unit) if and only if $\mathbb{E}$ is an M-normed vector lattice (with unit).

In the general case, the Moore distance $\delta$ may differ from the Hausdorff distance $\Delta$, as can be seen from [14; Example 6.12]. However, we have the following result which can also be used to give a simple proof of the properties of the Hausdorff distance on the class of all order intervals of an M-normed vector lattice with unit:

3.8. Lemma.

Suppose $\mathbb{E}$ is an M-normed vector lattice with unit.
Then the identity

$$\Delta([a,b],[c,d]) = \delta([a,b],[c,d])$$

holds for all order intervals $[a,b]$ and $[c,d]$ of $\mathbb{E}$.

For a proof of Lemma 3.8, see [14; Lemma 6.11].

Let $\mathbb{F}_b^{\Delta}(\mathbb{E}, |.|)$ denote the ordered cone $\mathbb{F}(\mathbb{E}, |.|)$ endowed with the Hausdorff distance $\Delta$.

3.9. Theorem.

Suppose $\mathbb{E}$ is a normed vector lattice.
Then $\mathbb{F}_b^{\Delta}(\mathbb{E}, |.|)$ is a normed ordered cone having the order cancellation property.
Moreover, $\mathbb{F}_b^{\Delta}(\mathbb{E}, |.|)$ is an M-normed semilattice cone with unit if and only if $\mathbb{E}$ is an M-normed vector lattice with unit.

For a proof of Theorem 3.9, see [14; Theorem 6.13 and Theorem 6.14]. We remark that in the case where $\mathbb{E}$ is an arbitrary normed vector lattice, $\mathbb{F}_b^{\Delta}(\mathbb{E}, |.|)$ may fail to be a normed semilattice cone since the Hausdorff distance need not be compatible with the semilattice structure of $\mathbb{F}_b(\mathbb{E}, |.|)$, as can be seen from [14; Example 6.10].

In the case where $\mathbb{E}$ is an arbitrary normed vector lattice of dimension greater than one, only the one-point sets of $\mathbb{E}$ are at the same time norm balls and order intervals of $\mathbb{E}$. However, in the case where $\mathbb{E}$ is even an M-normed vector lattice with unit, each norm ball of $\mathbb{E}$ is an order interval of $\mathbb{E}$ since there exists some $e \in \mathbb{E}_+$ satisfying $\overline{U} = [-e,e]$. In this case, $\mathbb{F}_b(\mathbb{E}, |.|)$ may be considered as the semilattice completion of $\mathbb{F}_b(\mathbb{E}, \|.\|)$, and the identity $\overline{U} = [-e,e]$ together with the role of the closed unit ball $\overline{U}$ in one of the equivalent definitions of the Hausdorff distance may also serve as an intuitive explanation of the compatibility of the Hausdorff distance with the semilattice structure of $\mathbb{F}_b(\mathbb{E}, |.|)$ .

## 4. REMARKS

As an abstraction of the structure of the class of all order intervals of an ordered vector space, Mayer [7] introduced the notion of a quasilinear space. Mayer also considered norms and metrics on a quasilinear space, but it appears that no ordered quasilinear spaces have been studied in the literature. This is somewhat surprising since ordered quasilinear spaces would reflect the inclusion of order intervals and would thus allow for a formulation of the subdistributive law for order intervals without any restriction on the scalars. On the other hand, with regard to the restricted distributive law in quasilinear spaces and the fact that a quasilinear space cannot be embedded into a vector space such that the embedding map is additive and homogeneous (and not only positively homogeneous), as pointed out by Kracht and Schröder [5], it seems to be convenient to generalize one step further and to restrict multiplication by scalars to positive scalars alone. This leads from quasilinear spaces to cones.

While the Minkowski addition of sets, the usual multiplication of sets by positive scalars, and the distances of Hausdorff and Moore are clearly related to the structure of the underlying vector space, this is not the case for the inclusion of sets. For order intervals of an ordered vector space $\mathbb{E}$ , a different order relation $\preccurlyeq$ can be defined by letting $[a,b] \preccurlyeq [c,d]$ if $a \leq c$ and $b \leq d$ holds; see Nickel [9] and Jahn [3]. The order relation $\preccurlyeq$ extends the order relation on $\mathbb{E}$ , and if $\mathbb{E}$ is even a vector lattice, then the order intervals of $\mathbb{E}$ form a lattice with respect to the order relation $\preccurlyeq$ , as pointed out by Nickel [9]. Let $\mathbb{F}_b(\mathbb{E}, \preccurlyeq)$ denote the class of all order intervals of $\mathbb{E}$ , endowed with the Minkowski addition of sets, the usual multiplication of sets by positive scalars, and the order relation $\preccurlyeq$ . Then $\mathbb{F}_b(\mathbb{E}, \preccurlyeq)$ is a lattice cone having the order cancellation property and the Riesz property but lacking the Hukuhara property. Furthermore, if $\mathbb{E}$ is a normed vector lattice, let $\mathbb{F}_b^\delta(\mathbb{E}, \preccurlyeq)$ denote the ordered cone $\mathbb{F}_b(\mathbb{E}, \preccurlyeq)$ endowed with the Moore distance $\delta$ . Then $\mathbb{F}_b^\delta(\mathbb{E}, \preccurlyeq)$ is a normed lattice cone. Since the cone properties of $\mathbb{F}_b(\mathbb{E}, \preccurlyeq)$ and $\mathbb{F}_b^\delta(\mathbb{E}, \preccurlyeq)$ are slightly different from those of $\mathbb{F}_b(\mathbb{E}, |.|)$ and $\mathbb{F}_b^\delta(\mathbb{E}, |.|)$ , they lead to different embedding theorems for order intervals reflecting the different properties of the order relation $\subseteq$ and $\preccurlyeq$ .

For the class of all order intervals of an (M-normed) vector lattice (with unit) $\mathbb{E}$ , it is also possible to prove embedding theorems which are concrete in the sense that the embedding vector lattice is defined to be the Cartesian product $\mathbb{E}\times\mathbb{E}$ , endowed with the componentwise defined addition, scalar multiplication, and order relation (and the sup-norm), and that the embedding map is defined by using the lower-bound-upper-bound representation of order intervals. These concrete embedding theorems can be obtained from an (isometrically) isomorphic representation of the (abstract) embedding vector lattice by $\mathbb{E}\times\mathbb{E}$ , and the different properties of the order relations $\subseteq$ and $<$ are then reflected by the different properties of the embedding map. Such concrete embedding theorems for $\mathbb{F}_b(\mathbb{E}, |.|)$ and $\mathbb{F}_b^{\delta}(\mathbb{E}, |.|)$ have been proven in [14]; see also [13] for the case $\mathbb{E} = \mathbb{R}$ .

Apart from the embedding theorems which can be obtained from the results of Sections 2 and 3, it appears that the investigation of the cone properties of classes of convex sets occuring in interval mathematics, which is motivated by Rådström's method for proving embedding theorems, may also be interesting for its own sake since it seems to be helpful for understanding the different properties of norm balls, hypernorm balls, and order intervals, and those of different order relations and metrics.

## REFERENCES

[ 1 ] Fischer, H.:
On the Rådström embedding theorem.
Analysis 5, 15-28 (1985).

[ 2 ] Hörmander, L.:
Sur la fonction d'appui des ensembles convexes dans un espace localement convexe.
Arkiv Mat. 3, 181-186 (1954-1958).

[ 3 ] Jahn, K.U.:
Maximale Fixpunkte von Intervallfunktionen.
Computing 33, 141-151 (1984).

[ 4 ] Kaucher, E.:
Algebraische Erweiterungen der Intervallrechnung unter Erhaltung der Ordnungs- und Verbandsstrukturen.
Computing Suppl. 1, 65-79 (1977).

[ 5 ] Kracht, M., and Schröder, G.:
Eine Einführung in die Theorie der quasilinearen Räume mit Anwendung auf die in der Intervallrechnung auftretenden Räume.
Math.-Phys. Semesterberichte Neue Folge 20, 226-242 (1973).

[ 6 ] Luxemburg, W.A.J., and Zaanen, A.C.:
Riesz Spaces I.
Amsterdam - New York - Oxford: North-Holland 1971.

[ 7 ] Mayer, O.:
Algebraische und metrische Strukturen in der Intervallrechnung und einige Anwendungen.
Computing 5, 144-162 (1970).

[ 8 ] Moore, R.E.:
Interval Analysis.
Englewood Cliffs, New Jersey: Prentice-Hall 1966.

[ 9 ] Nickel, K.:
Verbandstheoretische Grundlagen der Intervall-Mathematik.
In: Interval Mathematics.
Lecture Notes in Computer Science, vol. 29, pp. 251-262.
Berlin - Heidelberg - New York: Springer 1975.

[10] Rådström, H.:
An embedding theorem for spaces of convex sets.
Proc. Amer. Math. Soc. 3, 165-169 (1952).

[11] Ratschek, H.:
Nichtnumerische Aspekte der Intervallmathematik.
In: Interval Mathematics.
Lecture Notes in Computer Science, vol. 29, pp. 48-74.
Berlin - Heidelberg - New York: Springer 1975.

[12] Schaefer, H.H.:
Banach Lattices and Positive Operators.
Berlin - Heidelberg - New York: Springer 1974.

[13] Schmidt, K.D.:
Embedding theorems for classes of convex sets.
Acta Appl. Math. (to appear).

[14] Schmidt, K.D.:
Embedding theorems for classes of convex sets in a hypernormed vector space.
Analysis (to appear).

[15] Urbanski, R.:
A generalization of the Minkowski-Rådström-Hörmander theorem.
Bull. Acad. Polon. Sci. 24, 709-715 (1976).

# INTERVAL TEST AND EXISTENCE THEOREM

Shen Zuhe
Nanjing University
Nanjing
The People's Republic of China

## 1. Introduction

Interval methods have been introduced for computationally verifiable sufficient conditions for existence, uniqueness and convergence, for solving finite dimensional nonlinear systems [2], [7], [8], [9], [13] and for nonlinear operator equations in infinite dimensional spaces [11]. The methods can also be used to discuss some classical existence theorems [20], [21]. The conditions, like bounded inverse of the derivative, norm coercivity, or uniform monotonicity guarantee the homeomorphism of a differentiable function, and also, the nonsingularity assumption of the partial derivative guarantee the existence of a implicit function [14], [19]. In this paper, using interval methods, precisely, the centred form of the Newton-transform of f [3] and the Moore-like test for the Krawczyk operator [2], [8], we will derive some computationally verifiable sufficient conditions for a function f to be a homeomorphism and the global implicit function theorem. The cases of the function f(x) or f(x,y) to be local Lipschitz continuous and continuously differentiable are of special interest. Some further results for these classes of functions and a sufficient condition for the feasibility of the numerical continuation method are given.

## 2. Interval test

The Moore test [8] for a nonlinear system

$$f(x) = 0 \tag{1}$$

is based on the Krawczyk operator [2]

$$K(X) = y - \Gamma f(y) + \{I - \Gamma F'(X)\}(X - y).$$

The operator consists of the sum of the point iteration $\varphi(y) = y-\Gamma f(y)$ and a symmetric interval vector $\{I - \Gamma F'(X)\}(X - y)$. One can choose the real matrix $\Gamma$ for the Moore test. It is also possible to change the second part of the Krawczyk operator such that the Moore test becomes (at least sometimes!) more efficient. If we recognize that the Krawczyk operator is essentially the mean value form extension of $\varphi(y)$, then instead, other form extensions may also be used. This is indeed worth having and the centred form extension is perhaps a good choice and of practical value. The following example from [10] may illustrate the problem. Let

$$f(x) = \begin{pmatrix} x_1^2 + 0.25x_2 - 0{,}1725 \\ x_2^2 - 3x_1 + 0.46 \end{pmatrix}$$

$$X = \begin{pmatrix} [0,\ 0.5] \\ [0,\ 1] \end{pmatrix},\quad y = m(X) = \begin{pmatrix} 0.25 \\ 0.5 \end{pmatrix},\quad \Gamma = \begin{pmatrix} 0.8 & -0.2 \\ 2.4 & 0.4 \end{pmatrix}$$

we have [10]

$$K(X) = \begin{pmatrix} [0.03,\ 0.43] \\ [-0.02,\ 0.98] \end{pmatrix} \not\subset X.$$

By contrast, the centred form extension of $\varphi(x) = x - \Gamma f(x)$ on X is

$$K(X) = \begin{pmatrix} [0.1\ ,\ 0.33] \\ [0.23,\ 0.73] \end{pmatrix}.$$

It is clear that $K_c(X) \subset K(X)$ and $K_c(X) \subset X$. If the Moore test is true for $K_c(X)$ instead $K(X)$, it is then at least better to use $K_c(X)$ for this example.

The centred form extension was first suggested by Moore [7]. Explicit formulas for these forms were formed by Hansen [5] for polynomials and by Ratschek [18] for rational functions. Recursively defined centred forms were recently introduced by Krawczyk [4].

Let $\omega = (\omega_1, \omega_2, \cdots, \omega_n)$ be a symmetric interval, for an interval vector $X = (X_1, X_2, \cdots . X_n)$ and an interval matrix $A = (A_{ij})$, define [12]

$$\|X\|_\omega = \min\{\alpha \geqq 0 \mid X \subset \alpha\omega\}$$

$$\|A\|_\omega = \min\{\alpha \geqq 0 \mid A\omega \subset \alpha\omega\}$$

$$sp_\omega(X) = \|X - m(X)\|_\omega$$

and call them the interval norm, the interval matrix norm and the width of interval X respectively. It is easy to check the norm properties of a norm.

Suppose that $X \subset D \subset R^n$ is an interval, $f: D \subset R^n \to R^n$ is continuous on X and satisfies

$$f(x) - f(y) = h(x,y)(x-y), \quad x \in X, \; y \in R^n, \tag{2}$$

where $h(x,y) = (h_{ij}(x,y))$ and $h_{ij}(x,y): X \times R^n \to R^1$, $i,j = 1,2,\cdots,n$. For $n > 1$, there are in general infinitely many such matrices. Suppose that $h_{ij}(x,y)$ has an inclusion monotonic interval extension $H_{ij}(X,y)$, let

$$H(X,y) = (H_{ij}(X,y)), \tag{3}$$

and the Lipschitz condition

$$d(H(X,y), h(x,y)) \leqq \lambda d(X,y), \quad x \in X,$$

be satisfied. One possible such Lipschitz constant $\lambda$ is

$$\lambda = \frac{2\, sp_\omega(H(X,y))}{sp_\omega(X)}, \quad \text{for} \quad sp_\omega(X) > 0.$$

Define

$$F(X,y) = f(x) + H(X,y)(X - y), \quad y \in R^n$$

call it the centred form extension of f on X. For brevity, write $F(X) = F(X, m(X))$.

The following two special cases are of interest

(a) Suppose that $f(x): D \subset R^n \to R^n$ is local Lipschitz continuous on D [16]. For such maps, one can assign to each X, a certain collection $\partial f(x)$ of linear transformation from $R^n$ into $R^n$, called the generalized derivative and

$$\partial f(x) = \bigcap_{r>0} \overline{CO}\{f'(z) \mid f'(z) \text{ exists and } \|z - x\|_\infty < r\},$$

where $\overline{CO}$ is short for the closure of the convex hull.
Given $x_1, x_2 \in R^n$, if $[x_1, x_2] \subset D$, then we have the following Lipschitz mean value theorem

$$f(x_1) - f(x_2) = A(x_1 - x_2), \quad A \in \overline{CO}_{z \in [x_1, x_2]} \{\partial f(z)\}.$$

For each interval $X \subset D$, let

$$\partial f(X) = ([\underline{A_{ij}}, \overline{A_{ij}}]), \tag{4}$$

where

$$\underline{A_{ij}} = \min\{A_{ij} \mid A \in \overline{CO} \bigcup_{x \in X} \partial f(x)\},$$

$$\overline{A_{ij}} = \max\{A_{ij} \mid A \in \overline{CO} \bigcup_{x \in X} \partial f(x)\};$$

then $\partial f(X)$ is an inclusion monotonic interval extension of $\partial f(x)$ on X and

$$F(X) = f(m(X)) + \partial f(X)(X - m(X))$$

is a centred form extension of $f(x)$ on X.

(b) Let $f(x) = (f_1(x), f_2(x), \cdots, f_n(x)): D \subset R^n \to R^n$ with $f_i(x) \in C^1(D)$, then

$$f(x) - f(z) = f'(\xi(x,z))(x - z), \quad x, z \in D,$$

i.e., $h(x,z) = f'(\xi(x,z))$ can be defined. Let $F'(X)$ be an inclusion monotonic interval extension of $f'(x)$ on X, then

$$F(X) = f(z) + F'(X)(X - z)$$

is a centred form extension of f on X.

Krawczyk [3] also investigated the centred form extension of the Newton-transform and the corresponding interval operator for solving the equation (1), where $f: D \subset R^n \to R^n$ is continuous in the open set D. A simple way to convert the equation (1) to be solved into a fixed point problem is by introduction of the Newton-transform of f defined by

$$p(x) = x - \Gamma f(x),$$

where $\Gamma$ is a nonsingular real matrix. Let

$$r(x,y) = I - \Gamma h(x,y)$$

and

$$R(X,y) = I - \Gamma H(X,y),$$

where h(x,y) and H(X,y) are given in (2) and (3), respectively, then

$$P(X) = p(m(X)) + R(X)(X - m(X))$$

is the centred form extension of the Newton-transform p(x).

Theorem 1 [3].

(a) If $P(X) \subseteq X$, then there exists a fixed point $\hat{x} \in X$ of p(x);

(b) If $P(X) \subseteq \mathrm{Int}(X)$, then $\|R(X)\|_\omega \leqq \lambda < 1$.

## 3. The inverse function theorems (homeomorphism).

The inverse function is one of the most important notions in nonlinear analysis. The main theorem, for instance, is in the following ([14], Theorem 5.2.1, p.125):

Theorem 2.

Suppose that $f: D \subset R^n \to R^n$ has an F-derivative on a neighborhood of $x_0$, which is continuous at $x_0$ and that $f'(x_0)$ is nonsingular, then f is a local homeomorphism at $x_0$.

In this section, we will make use of interval methods to discuss the problem under weaker restrictions. In general, no differentiability of f on D is required.

Suppose that $f: D \subset R^n \to R^n$ is continuous on X and satisfies

$$f(x) - f(y) = h(x,y)(x - y),$$

where $h(x,y)$ is given in (2). Consider the Newton-transform of f

$$p(x) = x - \Gamma(f(x) - f(x_0))$$

and its centred form extension

$$P(X) = x_0 + (I - \Gamma H(X))(X - x_0), \tag{5}$$

where $H(X)$ is given in (3). We have

Theorem 3.

If for $x_0 \in X$, there exists an interval $X = [x_0 - \frac{\delta}{2}, x_0 + \frac{\delta}{2}]$ such that

$$P(X) \subseteq \text{Int}(X),$$

then $f(x)$ is a local homeomorphism at $x_0$.

Proof. The only thing we need to prove is the nonsingularity of $H(x)$ at $x_0$ (cf. [14], Theorem 5.1.9, p. 124). Since X is an interval, by Theorem 1, we have

$$\| I - \Gamma H(x) \| \leq \lambda < 1.$$

For brevitiy, we write $\| \ \|$ instead of $\| \ \|_\omega$ in the sequel. This proves that, for each $x \in X$, and of course, for $x_0$, $H(x)$ is nonsingular and so is $H(x_0)$.

□

Specially, suppose that $f: D \subset R^n \to R^n$ is local Lipschitz continuous on D. For $X \subset D$, define

$$H(X) = \partial f(X), \tag{6}$$

where $\partial f(X)$ is defined in (4), then we have

$$p(x) \in P(X)$$

and Theorem 3 is true for such functions.

Moreover, if f is continuously differentiable at $x_0$, define

$$H(X) = [\min_{x\in X} f'(x), \max_{x\in X} f'(x)],$$

then we have

$$p(x) \in P(X)$$

and the following stronger conclusion [20].

Theorem 4.

$f'(x)$ is nonsingular at $x_0$ if and only if there exists an interval $X = [x_0 - \frac{\delta}{2}, x_0 + \frac{\delta}{2}] \subset D$ such that

$$P(X) \subseteq Int(X). \tag{7}$$

Proof. We have only to prove the necessity. Let $\Gamma = [f'(x_0)]^{-1}$, from (5), we have

$$P(X) = x_0 + (I - \Gamma H(X))(X - x_0),$$

therefore, for any $x \in P(X)$, we have

$$x = x_0 + z,$$

where $z \in (I - \Gamma H(X))(X - x_0)$. Since

$$\begin{aligned} \|I - \Gamma H(X)\| &\leqq \|\Gamma\| \|f'(x_0) - F'(X)\| \\ &= \|\Gamma\| \|f'(x_0) - [\min_{x\in X} f'(x), \max_{x\in X} f'(x)]\| \end{aligned}$$

and f is continuous at $x_0$, there exists $\delta > 0$ such that, for $X = [x_0 - \frac{\delta}{2}, x_0 + \frac{\delta}{2}]$, $\|I - \Gamma H(X)\| < 1$ and $\|(I - \Gamma H(X))(X - x_0)\| < \frac{\delta}{2}$, therefore

$$\|P(X) - x_0\| < \frac{\delta}{2},$$

and (7) is proved.

□

Corollary 5.

If for $x_0 \in X^{(0)}$, there exists an interval $X = [x_0 - \frac{\delta}{2}, x_0 + \frac{\delta}{2}] \subset X^{(0)} \subset D$ such that

$$P(X) \subseteq \text{Int}(X),$$

then $f(x)$ is a local homeomorphism at $x_0$.

## 4. The global inverse function theorem.

Following the above discussion, we will give a computationally verifiable sufficient condition for the global inverse function theorem. In this way, the theory of continuation is required.

Definition 1.

The continuous mapping $f: D \subset R^n \to R^n$ satisfies condition (L) if for every $(x_0, y) \in D \times R^n$ and every path $q: [0,b) \to D$, $b \leq 1$ such that $f(q(t)) = (1-t)f(x_0) + ty$, $0 \leq t < b$, there exists a sequence $t_n \to b$ as $n \to \infty$ such that $\lim_{n\to\infty} q(t_n)$ exists and is in D.

Theorem 6 (Plastock [15]).

Let $f: D \subset R^n \to R^n$ be a local homeomorphism at each point $x \in D$. The condition (L) is necessary and sufficient for $f(x)$ to be a homeomorphism from D onto $R^n$.

In connection with interval methods, we have

Theorem 7.

If for each $x \in D$, there exist an interval $\overline{X}_x = [x - \frac{\delta(x)}{2}, x + \frac{\delta(x)}{2}]$ and a nonsingular real matrix $\Gamma(x)$, continuous in x such that

(a) $\|\Gamma(x)\| \leqq M$, M = constant, $x \in D$;

(b) there exists a constant $\alpha$, $0 \leqq \alpha < 1$ such that

$$|P(X_x) - m(X_x)| \leqq \alpha \frac{(x)}{2} ,$$

then f(x) is a homeomorphism from D onto $R^n$.

Proof. By (b), f(x) is a local homeomorphism at each point $x \in D$. The only problem remained to prove is that f(x) satisfies condition (L). It follows that, for each $x \in X_x$, and consequently, for each $x \in D$, we have

$$\|I - \Gamma H(x)\| \leqq \alpha < 1,$$

therefore

$$\|H(x)^{-1}\| \leqq \frac{\|\Gamma(x)\|}{1-\alpha} \leqq \frac{M}{1-\alpha} .$$

Now, for every $(x_0,y) \in D \times R^n$ and every path $q(t): [0,b) \to D$ such that

$$f(q(t)) = (1-t)f(x_0) + ty = p(t), \quad 0 \leqq t < b ,$$

we have

$$f(q(t_1)) - f(q(t_2)) = p(t_1) - p(t_2)$$

$$= (t_1 - t_2)(y - f(x_0)), \quad t_1,t_2 \in [0,b),$$

and

$$F(q(t_1)) - F(q(t_2)) = H(q(t_1),q(t_2))(q(t_1) - q(t_2)).$$

It is clear that $H(q(t_1),q(t_2))$ is also nonsingular and

$$\|H(q(t_1),q(t_2))^{-1}\| \leqq \frac{M}{1-\alpha} , \text{ for any } t_1,t_2 \in [0,b),$$

therefore

$$\|q(t_1) - q(t_2)\| \leqq \frac{M}{1-\alpha} \|y - f(x_0)\| \, |t_1 - t_2|.$$

It means that $q(t)$ is Lipschitz and $\lim_{t \to b} q(t)$ exists. This proves that $f$ is a homeomorphism from $D$ onto $R^n$. □

Corollary 8.

If $f(x)$ is local Lipschitz continuous on $D$ and define $H(X)$ as in (6), then the conditions (a) and (b) in Theorem 7 are sufficient for $f(x)$ to be a homeomorphism from $D$ onto $R^n$.

The other special case is that $f(x)$ is continuously differentiable on $D$. Let $\Omega$ be the set of all continuous mappings $\omega$: $R_+ \to R_+$ satisfying the following properties

$$\omega(t) > 0 \quad \text{for} \quad t > 0 \quad \text{and} \quad \int_1^\infty \frac{dt}{\omega(t)} = \infty.$$

Theorem 9 (M. and S. Radulescu [17] and Mihai Cristea [6]).

Let $a,b \in R^n$, $a < b$ and $\sigma = \{x \in R^n,\ a \leqq x < b\}$, $u,v$: $\sigma \to R_+$, $u$ be continuous and $v$ integrable on $\sigma$. If $c \geqq 0$ and $\omega \in \Omega$ is a map with the properties

$$\int_a^b v(y)\omega(u(y))dy < \infty$$

$$u(x) \leqq c + \int_a^x v(y)\omega(u(y))dy\ , \qquad x \in \sigma,$$

then the following inequality holds

$$\int_0^{u(x)} \frac{ds}{\omega(s)} \leqq \int_0^x v(y)dy.$$

With Theorem 9, we have

Theorem 10.

Let $\omega(x) \in \Omega$, $f$: $D \subset R^n \to R^n$ be a continuously differentiable map. If for each $x \in D$, there exist an interval $X_x$ with the width $\delta(x) > 0$, $x \in X_x$ and a nonsingular real matrix $\Gamma(x)$ such that

(a) $\|\Gamma(x)\| \leqq \omega(\|x\|)$, $x \in D$;

(b) there exists a constant $\alpha$, $0 \leqq \alpha < 1$ such that

$$|P(X_x) - m(X_x)| \leqq \alpha \frac{\delta(x)}{2},$$

then $f(x)$ is a homeomorphism from D onto $R^n$.

Proof. By the condition of the theorem, $f(x)$ is a local homeomorphism at each point x in D and $f'(x)$ is nonsingular with

$$\|f'(x)^{-1}\| \leqq \frac{1}{1-\alpha}\|\Gamma(x)\| \leqq \frac{\omega(\|x\|)}{1-\alpha}, \quad x \in D.$$

Now, for every $x_0, y \in R^n$, $p(t) = (1-t)f(x_0) + ty$, $0 \leqq t < 1$, $q(t)\colon [0,b) \to R^n$, $b \leqq 1$ such that

$$q(0) = x_0, \quad f(q(t)) = p(t), \quad t \in [0,b), \tag{8}$$

we may assume that the path $q(t) \in C^1$. Differentiating the equality (8), we obtain

$$f'(q(t))q'(t) = p'(t),$$

whence $q'(t) = (f'(q(t))^{-1}(y - f(x_0))$, $t \in [0,b)$. Since

$$\begin{aligned} \|q(t) - q(0)\| &\leqq \int_0^t \|q'(s)\| ds \leqq \int_0^t \|f'(q(s))^{-1}\| \|p'(s)\| ds \\ &\leqq \int_0^t \frac{\omega(\|q(s)\|)}{1-\alpha} \|p'(s)\| ds \\ &= \frac{\|f(x_0) - y\|}{1-\alpha} \int_0^t \omega(\|q(s)\|) ds, \end{aligned}$$

$$0 \leqq t < b,$$

we have

$$\|q(t)\| \leqq \|q(0)\| + \frac{\|f(x_0) - y\|}{1-\alpha} \int_0^t \omega(\|q(s)\|) ds, \qquad 0 \leqq t < b.$$

By applying Theorem 9 to the preceeding inequality, it follows that

$$\int_{\|q(0)\|}^{\|q(t)\|} \frac{ds}{\omega(s)} \leq \int_0^t \frac{\|f(x_0) - y\|}{1-\alpha} ds ,$$

whence there exists $k > 0$ such that $\|q(t)\| \leq k$, $0 \leq t < b$. We see that $0 \leq t_1, t_2 < b$ implies

$$\|q(t_1) - q(t_2)\| \leq \int_{t_1}^{t_2} \|q'(s)\| ds \leq \int_{t_1}^{t_2} \|f'(q(s))^{-1}\| \|f(x_0)-y\| ds$$

$$< \frac{\|f(x_0) - y\|}{1-\alpha} \omega(k) |t_1 - t_2| .$$

therefore q is Lipschitz. Consequently, it follows that $\lim_{t \to b} q(t)$ exists. Thus, we have proven that f satisfies condition (L). □

Note. The assumption "$\Gamma(x)$ is uniformly continuous on D or $\Gamma(x)$ is a constant nonsingular matrix" is sufficient for the condition (a) in Theorem 10.

## 5. The feasibility of the numerical continuation method.

Many iterative methods will converge to a solution $x^*$ of $f(x) = 0$ only if the initial approximations are sufficiently close to $x^k$. The continuation methods may be considered as an attempt to widen this domain of convergence of a given method or as a procedure to obtain sufficiently close starting points. The method can be described as follows: Let $f: D \subset R^n \to R^n$ be a given mapping and consider the problem of solving the equations (1). This problem is imbedded into a family of problems of the form

$$G(t,x) = 0, \tag{9}$$

where $t \in [0,1]$ is a parameter. The imbedding is chosen so that at $t = 0$, the solution of (9) is a known point $x_0$, while at $t = 1$, the solution $x_1$ of (9) also solves (1). For example, $G(t,x)$ might have the form

$$G(t,x) = tf(x) + (1-t)f(x_0). \tag{10}$$

Suppose now that there exists a continuous solution curve $x(t): [0,1] \to D$ of (9) starting at $x_0$, the continuation method then involves proceeding in some as yet unspecified manner along or near this curve $x = x(t)$ from the initial point $x_0 = x(0)$ to the final $x_1 = x(1)$. For each t, consider the iterative process

$$x^{n+1} = G(t,x^n).$$

In general, this process will converge to x(t) only for starting values near that point, and we cannot hope that we have convergence for the process with t = 1 starting from x(0). This lead to the following numerical continuation process: a partition of [0,1]

$$0 = t_0 < t_1 < \cdots < t_n = 1 \tag{11}$$

and a sequence of integers $\{j_k\}$, $k = 1,2,\cdots,N-1$ is chosen such that the points

$$\begin{aligned} x_k^{j+1} &= G(t_k,x_k^j), \quad j = 0,1,\cdots,j_{k-1}, \; k = 1,2,\cdots,N-1 \;, \\ x_{k+1}^0 &= x_k^{j_k} \;, \quad x_1^0 = x(0) \end{aligned} \tag{12}$$

are well-defined and such that

$$x_N^{j+1} = G(1, x_N^j), \quad j = 0,1,2,\cdots \tag{13}$$

converges to x(1) as $j \to \infty$.

Although many authors have discussed the numerical continuation approach, little attention appears to have been paid to its feasibility. Avila [1] established the feasibility of the numerical continuation process under nearly minimal conditions.

<u>Definition 2</u>.

If a partition (11) exists so that with some sequence of integers $\{j_n\}$ the entire process (12), (13) is well-defined and so that (13) converges to x(1), then the numerical continuation process (12), (13) is called feasible.

Theorem 11 (Avila [1]).

Let $G: [0,1]\times D \subset [0,1]\times R^n \to R^n$. Assume that $x: [0,1] \to D$ is continuous and satisfies $x(t) \equiv G(t,x(t))$.
Let G have a strong partial derivative with respect to x at $(t,x(t))$ for every $t \in [0,1]$. If the spectral radius of $G_x(t,x(t))$

$$\rho(G_x(t,x(t))) < 1$$

for all $t \in [0,1]$, then the numerical continuation process (12), (13) is feasible.

Two basic assumptions are involved in Avila's theorem: The existence of the solution curve $x(t)$ and the spectral radius $\rho(G_x(t,x(t)) < 1$ for all $t \in [0,1]$, which are certainly difficult to test. In this section, we will also make use of interval analysis to give simpler sufficient conditions for Avila's theorem.

Let D be an open set, $f: D \subset R^n \to R^n$ be continuously differentiable on D. If for each $x \in D$, there exists an interval $X_x$ with the width $\delta(x)$, $x \in X_x$ and a nonsingular real matrix $\Gamma(x)$, continuous in x such that the conditions (a), (b) in Theorem 10 are satisfied, then by Theorem 10 f is homeomorphism from D onto $R^n$. That is, for each $t \in [0,1]$, there exists a unique point $x(t) \in D$ such that

$$G(t,\gamma(t)) = tf(x(t))+(1-t)f(x_0) = 0$$

and $x(t)$ is continuous on $[0,1]$. At the same time, for each $t \in [0,1]$ and for the interval mapping

$$P_{x(t)}(X) = m(X)-\Gamma(x(t))(tf(x(t))+(1-t)f(x_0))+(I-\Gamma(x(t))H(X))(X-m(X)),$$

there exists an interval $X_{x(t)}$ with the width $\delta(x(t))$ such that

$$|P_{x(t)}(X_{x(t)})-m(X_{x(t)})| \leqq \alpha \, \frac{\delta(x(t))}{2}$$

and consequently

$$\|I - \Gamma(x(t))H(X_{x(t)})\| \leqq \alpha < 1, \qquad (14)$$

therefore, for each t, the sequence

$$x^{k+1}(t) = x^k(t) - \Gamma(x^k(t))G(t,x^k(t)) \tag{15}$$

converges to the unique solution x(t) for any starting point $x^0(t)$ in $X_{x(t)}$.

We now turn our attention to the Newton-transform (15) as the local process in numerical continuation method. Consider the mapping $p(t,x)$: $[0,1]\times D \rightarrow R^n$ given by

$$p(t,x) = x - \Gamma(x)G(t,x).$$

For a certain partition $\{t_k\}$, $k = 0,1,\cdots,N$ of $[0,1]$ and integers $\{j_k\}$ $k = 1,2,\cdots,N-1$, the process is defined by the equations

$$\begin{aligned} x_k^{j+1} &= x_k^j - \Gamma(x(t_j))G(t_j,x_k^j),\ k = 1,2,\cdots,N-1,\ j=0,1,\cdots,j_{k-1}, \\ x_{k+1}^0 &= x_k^{j_k},\quad x_1^0 = x(0) \end{aligned} \tag{16}$$

followed by

$$x_N^{j+1} = x_N^j - \Gamma(x(1))G(1,x_N^j),\quad x_N^0 = x_{N-1}^{j_{N-1}},\ j = 0,1,\cdots \tag{17}$$

then we have

Theorem 12.

Under the condition (a), (b) in Theorem 10, the numerical continuation method (16), (17) is feasible.

Proof. In fact, by the Lemma 3.1 in [1], we see that

$$G_x(t,x(t)) = I - \Gamma(x(t))f'(x(t))$$

and $G_x(t,x)$ is continuous in $(t,x)$. Moreover, from (14), we have

$$\rho(G_x(t,x(t)) \leqq \alpha < 1,\ t \in [0,1].$$

Thus, the conditions in Theorem 11 are satisfied and the feasibility of (16), (17) is proved.

## 6. The implicit function theorem.

The implicit function is also an important problem in nonlinear analysis. We refer to the book [14], Theorem 5.2.4 for the general result of the theorem. In this section, the interval method will also be used to discuss this problem under weaker restrictions.

Suppose that $f(x,y):==D_1\times D_2\subset R^n\times R^n\to R^n$ is continuous on an open set D and there is an identity

$$f(x,y_1) = f(x,y)+h(x,y_1,y)(y_1-y), \quad x\in D_1,\ y_1,y\in D_2,$$

where $h(x,y_1,y) = (h(x,y_1,y)_{ij})$.
Suppose that $H(x,Y,y)$ is an inclusion monotonic interval extension of $h(x,y_1,y)$ on Y, $Y\subset Y^{(0)}\subset D_2$ with the Lipschitz condition

$$d(H(x,Y,y),\ h(x,y_1,y)) \leqq \lambda\, d_x(Y,y).$$

For brevity, write $H(x,Y,m(Y)) = H(x,Y)$. Consider an interval $X^{(0)}\times Y^{(0)}\subset D$ and a mapping

$$p_x(y) = y-\Gamma f(x,y),$$

where $\Gamma$ is a nonsingular real matrix, then for all $x\in X^{(0)}$, $p_x(y)$ is continuous on $Y^{(0)}$ and

$$p_x(y)\in P_x(Y), \quad y\in Y,$$

where

$$P_x(Y) = m(Y)-\Gamma f(x,m(Y))+(I-\Gamma H(x,Y))(Y-m(Y)), \tag{18}$$

Theorem 13 [20].
If for fixed $x^0\in X\subset X^{(0)}$, there exists an interval $Y\subset Y^{(0)}$ such that

$$P_{x^0}(Y) \leqq \mathrm{Int}(Y),$$

then

(a) $f(x^0,y) = 0$ has a unique solution $y^0 = y(x^0) \in Y$;

(b) for each $y \in Y$, $H(x^0,y)$ is nonsingular;

(c) there exists an interval $X = [x^0 - \delta_1, x_0 + \delta_1]$ and an interval $Y = [y^0 - \delta_2, y^0 + \delta_2] \subset Y^0$ such that, for any $x \in X$, $f(x,y) = 0$ has a unique solution $y = y(x) \in Y$ and the mapping $y: X \to Y$ is continuous.

Proof. We can apply the preceeding results in the section 3 to prove (a) and (b). (c) follows immediately from (a), (b) and Theorem 5.2.4 in [14].

Continuating the result, we will also have a computationally verifiable sufficient condition for the global implicit function theorem. The theorem has the special feature that it does not require an initial solution point of the implicit function.

Theorem 14.

If for each $x \in D_1$ and every $y \in D_2$, there exists an interval $Y_x = [y - \frac{\delta_x(y)}{2}, y + \frac{\delta_x(y)}{2}]$ and a nonsingular real matrix $\Gamma_x(y)$, continuous in y, such that

(a) $\|\Gamma_x(y)\| \leqq M$, $M = \text{constant}$;

(b) there exists a constant $\alpha(x)$, $0 \leqq \alpha(x) < 1$ such that

$$|P_x(Y_x) - m(Y_x)| \leqq \alpha(x) \frac{\delta_x(y)}{2},$$

then for each $x \in D_1$, there exists a unique solution $y(x)$ to $f(x,y) = 0$ and $y(x)$ is continuous on $D_2$.

The proof of this theorem is similar to that of Theorem 7, we do not give it here. □

Similarly, the following two special cases are interested.

(a) Suppose that $f(x,y)$ is local Lipschitz with respect to y. For each $x \in X^{(0)}$ and interval $Y \subset Y^{(0)}$, let

$$\partial_2 F(x,Y) = [\underline{A_{ij}}(x), \overline{A_{ij}}(x)],$$

where

$$\underline{A_{ij}}(x) = \min \{ A_{ij}(x) \mid A \in \overline{CO} \bigcup_{y \in Y} \partial_2 f(x,y) \},$$

$$\overline{A_{ij}}(x) = \max \{ A_{ij}(x) \mid A \in \overline{CO} \bigcup_{y \in Y} \partial_2 f(x,y) \}.$$

and $\partial_2 f(x,y)$ is the generalized derivative with respect to y. Let $P_x(Y)$ be in (18) with $H(x,Y) = \partial_2 f(x,Y)$.

(b) Suppose that f(x,y) and its derivative $f_y(x,y)$ are continuous on D. Define

$$f_y(x,Y) = [\min_{y \in Y} f(x,y), \max_{y \in Y} f(x,y)]$$

and let $P_x(Y)$ be given as in (18) with $H(x,Y) = f_y(x,Y)$.

Corollary 15.

In both cases (a) and (b), Theorem 14 is valid.

Specially, if $f_x(x,y)$ exists at $(x^0,y^0)$ and case (b) applies, then y(x) is differentiable at $x^0$ and

$$y'(x^0) = -[f_y(x^0,y^0)]^{-1} f_x(x^0,y^0).$$

Furthermore, by the differentiability assumption of f(x,y), the following converse conclusion can be proved.

Theorem 16.

Suppose that $f(x,y): D \subset R^n \times R^n \to R^n$ is continuous on an open neighborhood of $(x^0,y^0)$ for which $f(x^0,y^0) = 0$. Assume that $f_y(x,y)$ exists in a neighborhood of $(x^0,y^0)$ and is continuous at $(x^0,y^0)$ and that $f_y(x^0,y^0)$ is nonsingular, then there exists an interval $Y = [y^0 - \frac{\delta}{2}, y^0 + \frac{\delta}{2}]$ such that

$$P_{x^0}(Y) \subseteq \mathrm{Int}(Y). \qquad (19)$$

Proof. Let $\Gamma = [f_y(x^0,y^0)]^{-1}$, then

$$P_{x^0}(Y) = y(x^0) - \Gamma f(x^0,y(x^0)) + (I - \Gamma f(x^0,Y))(Y - y(x^0)).$$

For any $y \in P_{x^0}(Y)$, we have

$$y = \beta + \mu$$

with $\beta = y(x^0) - \Gamma f(x^0,y(x^0)) = y(x^0)$ and $\mu \in (I - \Gamma fy(x^0,Y))(Y - y(x^0))$.
Since

$$\begin{aligned} \|I - \Gamma f_y(x^0,Y)\| &\leq \|\Gamma(f_y(x^0,y^0) - f_y(x^0,Y))\| \\ &\leq \|\Gamma\|\|f_y(x^0,y^0) - [\inf_{y\in Y} f_y(x^0,y), \sup_{y\in Y} f_y(x^0,y)]\| \\ &\leq \|\Gamma\|\max\{\inf_{y\in Y}(f_y(x^0,y^0) - f_y(x^0y)), \\ &\qquad \sup_{y\in Y}(f_y(x^0,y^0) - f_y(x^0,y))\} \end{aligned}$$

and by the continuity of $f_y(x,y)$ on D, there exists $\delta$, such that for $y \in Y = [y^0 - \frac{\delta}{2}, y^0 + \frac{\delta}{2}]$, we have

$$\|I - \Gamma f_y(x^0,Y)\| < 1$$

and

$$\|(I - \Gamma f_y(x^0,Y))(Y - y^0)\| < \frac{\delta}{2},$$

therefore

$$\|P_{x^0}(Y) - y^0\| < \frac{\delta}{2}$$

and (18) is proved.

□

Under the differentiability assumption of f(x,y), we can also obtain a stronger global implicit function theorem.

Theorem 17.

If for each fixed $x \in D$, and every $y \in D_2$, there exists an interval $Y_x = [y - \frac{\delta_x(y)}{2}, y + \frac{\delta_x(y)}{2}]$ and a nonsingular real matrix $\Gamma_x(y)$, continuous in y such that

(a) $\|\Gamma_x(y)\| < \omega_x(\|y\|)$, $y \in D_2$, $\omega_x(t) \in \Omega$;

(b) there exists a constant $\alpha(x)$, $0 \leq \alpha(x) < 1$ such that

$$|P_x(Y_x) - m(Y_x)| < \alpha(x)\, \frac{\delta_x(y)}{2},$$

then

(a) for each $x \in D_1$, there exists a unique solution $y(x)$ to $f(x,y) = 0$ and $y(x)$ is continuous on $D_1$;

(b) if $f_x(x,y)$ exists and is continuous on D, then

$$y'(x) = -[f_y(x,y)]^{-1} f_x(x,y).$$

In connection with the theorem of M. and S. Radulescu [17], we have the following sufficient condition for Theorem 17.

Theorem 18.

Let $\omega \in \Omega$, $f(x,y): D = D_1 \times D_2 \subset R^n \times R^n \to R^n$ be continuously differentiable on D and for all $(x,y) \in D$, and

(a) $f_y(x,y)$ be an isomorphism from $D_2$ on $D_2$;

(b) $\|f_y(x,y)^{-1}\| (1 + \|f_x(x,y)\|) \leq \omega(\max(\|x\|, \|y\|))$,

then the condition of Theorem 17 will be hold with

$$\omega_x(\|y\|) = \begin{cases} \omega(\|x\|), & \text{for } \|x\| \geq \|y\|; \\ \omega(\|y\|), & \text{for } \|y\| \geq \|x\|. \end{cases}$$

Since the proofs of Theorem 17 and 18 are similar to those of Theorem 10 and 16, respectively, we do not give them, too.

## 7. Acknowledgments

I wish to express my deep gratitute to Prof. Dr. K. Nickel for his financial support which made it possible for me to participate the International Interval Symposium in Freiburg, 1985.

I am also thankful to Professors K. Nickel, R.E. Moore, Dr. A. Neumaier and Dr. J. Garloff for their invaluable help and advices, and also to Mrs. L. Wüst for her excellent typewritting.

## References

[1] Avila, J.H.: The feasibility of continuation methods for nonlinear equations, SIAM. J. Numer. Anal. 11 (1974).

[2] Krawczyk, R.: Newton-Algorithmen zur Bestimmung von Nullstellen mit Fehlerschranken, Computing 4, 187-201 (1969).

[3] Krawczyk, R.: Zentrische Formen und Intervalloperatoren, Freiburger Intervall-Berichte 82/1, 1-30 (1982).

[4] Krawczyk, R.: Intervallsteigungen für rationale Funktionen und zugeordnete zentrische Formen. Freiburger Intervall-Berichte 83/2, 1-30 (1983).

[5] Hansen, E.R.: The centred form. in: Topics in Interval Analysis, (E.R. Hansen, Ed.) Oxford, pp. 102-105.

[6] Mihai Cristea: A note on global inversion theorems and applications to differential equations. Nonlinear Analysis, Theory, Methods and Applic., pp. 1155-1161 (1981).

[7] Moore, R.E.: Interval Analysis. Prentice-Hall, Englewood Cliffs, NJ (1966).

[8] Moore, R.E.: A test for existence of solution to nonlinear systems, SIAM J. Numer. Anal. 14, 611-615 (1977).

[9] Moore, R.E.: Methods and Applications of Interval Analysis. SIAM Publications, Philadelphia (1974).

[10] Moore, R.E., and Qi, L.: A successive interval test for nonlinear systems, SIAM J. Numer. Anal. 19 (1982).

[11] Moore, R.E., and Zuhe Shen: An interval version of Chebyshev's method for nonlinear operator equations. Nonlinear Analysis, Theory, Methods and Applic. 7, 21-34 (1983).

[12] Neumaier, A.: Interval norms, Freiburger Intervall-Berichte 81/5, 17-33 (1981).

[13] Nickel, K.: On the Newton method in interval analysis. MRC Report No. 1136, Mathematics Research Center. University of Wisconsin-Madison, (1971).

[14] Ortega, J.M., and Rheinboldt, W.C.: Iterative Solution of Non-linear Equations in Several Variables, Academic Press, New York-London (1970).

[15] Plastock. R.: Homeomorphism between Banach spaces. Trans. Amer. Math. Soc., 200, 169-183 (1974).

[16] Pourciau, B.H.: Analysis and optimization of Lipschitz continuous mappings. J. Optim. Theory Appl. 22, 311-351 (1977).

[17] Radulescu, M., and Radulescu, S.: Global inversion theorem and applications to differential equations. Nonlinear Analysis, Theory, Methods and Applic. 4, 951-965 (1980).

[18] Ratschek, H.: Zentrische Formen. ZAMM 58, 434-436 (1978).

[19] Schwetlick, H.: Numerische Lösung nichtlinearer Gleichungen, VEB Deutscher Verlag des Wissenschaften, Berlin, 1978.

[20] Zuhe, Shen: On some classical existence theorems. Nonlinear Analysis, Theory, Methods and Applic. 7, 1024-1033 (1983).

[21] Zuhe, Shen: Diffeomorphism and the feasibility of the numerical continuation methods. Nonlinear Analysis, Theory, Methods and Applics. 9, 495-502 (1985).

# TECHNICAL CALCULATIONS

# BY MEANS OF INTERVAL MATHEMATICS

P. Thieler

Fachhochschule Darmstadt

D-6100 Darmstadt

## Motivation

The idea of this paper is to add some elementary examples of applications to the wide spread and very few others that seem to exist in the interval mathematics literature. The examples presented here are intended to motivate engineer students very early to deal with interval mathematics. They can be inserted into the first lessons of a lecture on calculus, e.g. after having introduced inequalities ( and interval arithmetic ! ), as well as into the exercise collection of a programming course for all sorts of students that need mathematics as an instrument of practical support in their future professions.

Note that the technical surroundings of the examples are the bridge to the engineer student's ear, whereas the simplicity of the formulas dealt with is the way to surround his fear : Interval mathematics are useful a n d easy !

## Interval Arithmetic in HP BASIC

All printed programs or numerical results are produced on the desk top computers HEWLETT PACKARD 9845T of the Mathematisches Labor I, Fachhochschule Darmstadt. The programming language used here is a very high level structured BASIC which is at least as powerful as FORTRAN 77. It is combined with a BASIC-written precompiler of Heinz LERCHE and the author and completed by a BASIC interval arithmetic package developed by the students Marika GAUCH, Bernd GUTHIER, Gerhard HORNBERGS and Reiner UHL and the author. For more information, see /T/.

## Electric Current Rules

When studying the elements of electricity, one is faced to many simple formulas which can easily ( and should be ) written in terms of interval arithmetic. The first example shows how to examplify all f o u r e l e m e n t a r y o p e r a t i o n s at once and how to demonstrate the practical handling of c o n s t a n t s and n-th p o w e r s.

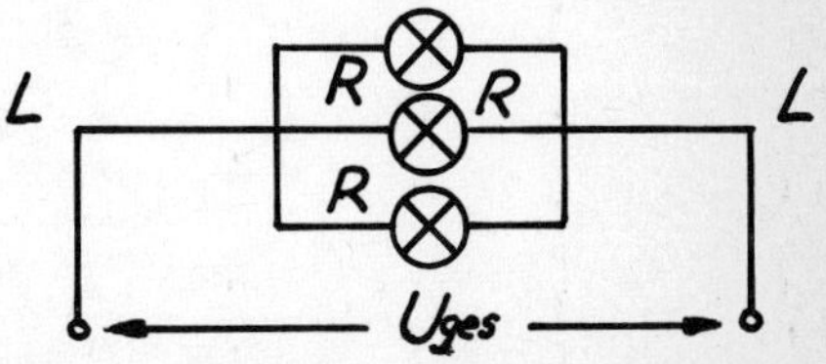


Problem. Three bulbs of R=240Ω resistance each are connected by a L=100m long aluminium line of diameter d=1.5mm and specific resistance $\varrho$=0.02857Ωmm$^2$/m to a source Uges=220V of electromotive force. The accuracy of all data is ± 0.5 percent. - Switch on one, two or three lamps. Which is the operating voltage in each case ? (cf. /L/,Nr.947)

Solution. (a) One lamp :

| | | |
|---|---|---|
| Resistance | : | Rges = Rl + R , |
| current | : | Iges = Uges / Rges , [1] |
| voltage drop in the line | : | Ul = Rl * Iges , |
| operating voltage (1 bulb): | : | U1 = Uges - Ul . |

First calculate $\lfloor Rl \rfloor$ in consideration of all possible errors. Then define

$$\lfloor R \rfloor := [\ 238.8, 241.2\ ]\ ,\ \lfloor Uges \rfloor := [\ 218.9, 221.1\ ]$$

and perform

$$\lfloor Rges \rfloor = \lfloor Rl \rfloor \oplus \lfloor R \rfloor ,$$
$$\lfloor Iges \rfloor = \lfloor Uges \rfloor \oslash \lfloor Rges \rfloor ,$$
$$\lfloor Ul \rfloor = \lfloor Rl \rfloor \circledast \lfloor Iges \rfloor ,$$
$$\lfloor U1 \rfloor = \lfloor Uges \rfloor \ominus \lfloor Ul \rfloor .$$

(The ideal operations $\oplus$ , ... , $\oslash$ will later be replaced by the corresponding machine operations that provide rounded interval arithmetic.)

The determination of the line resistance Rl is done as follows :

| | | |
|---|---|---|
| Line length | : | $Ll = 2 * L$ , |
| sectional area of line | : | $A = \pi * ( d / 2 )^2$ , |
| line resistance | : | $Rl = ( \varrho * Ll ) / A$ |
| | | $= \varrho * L * 8 / ( \pi * d^2 )$ . |

Introduce the interval constant $0.0\underset{\bullet}{2}857$ of width zero and execute

$$\lfloor Rho \rfloor = 0.0\underset{\bullet}{2}857 \circledast [\ 0.995, 1.005\ ]$$

to fix the possible error of $\varrho$ . After the definition of $\underset{\bullet}{8}$ and $\lfloor \pi \rfloor$ , one gets

$$\lfloor Rl \rfloor = \lfloor Rho \rfloor \circledast \lfloor L \rfloor \circledast \underset{\bullet}{8} \oslash ( \lfloor \pi \rfloor \circledast \lfloor d \rfloor \circledast \lfloor d \rfloor )$$

with $\lfloor d \rfloor := 1.\underset{\bullet}{5} \circledast [\ 0.995, 1.005\ ]$ .

(b) Two lamps : The equation
Uges = Ul + U12 = Rl * Iges + R * ( Iges / 2 ) = ( Rl + R / 2 ) * Iges
leads to

$$\lfloor Rges \rfloor = \lfloor Rl \rfloor \oplus \lfloor R \rfloor \oslash \underset{\bullet}{2}$$

which replaces the corresponding formula in (a).

(c) Three lamps : Since
Uges = Ul + U123 = Rl * Iges + R * ( Iges / 3 ) = ( Rl + R / 3 ) * Iges ,
now

$$\lfloor Rges \rfloor = \lfloor Rl \rfloor \oplus \lfloor R \rfloor \oslash \underset{\bullet}{3}$$

holds.

(d) One to three lamps : All cases can be treated s i m u l t a n e o u s l y when using interval arithmetic :

$$\lfloor Rges \rfloor = \lfloor Rl \rfloor \oplus \lfloor R \rfloor \circledast [\ 1/3\ ,\ 1\ ] .$$

Numerical results.

| | | |
|---|---|---|
| R | [ 2.38799999999E+02 | , 2.41200000001E+02 ] |
| L | [ 9.94999999999E+01 | , 1.00500000001E+02 ] |
| D | [ 1.49249999999E+00 | , 1.50750000001E+00 ] |
| Rho | [ 2.84271499999E-02 | , 2.87128500001E-02 ] |
| Uges | [ 2.18899999999E+02 | , 2.21100000001E+02 ] |
| Rl | [ 3.16943518253E+00 | , 3.29878338595E+00 ] |
| Rges | [ 2.41969435181E+02 | , 2.44498783388E+02 ] |
| Iges | [ 8.95300978457E-01 | , 9.13751771317E-01 ] |
| Ul | [ 2.83759842007E+00 | , 3.01426916211E+00 ] |
| U1 | [ 2.15885730836E+02 | , 2.18262401582E+02 ] |

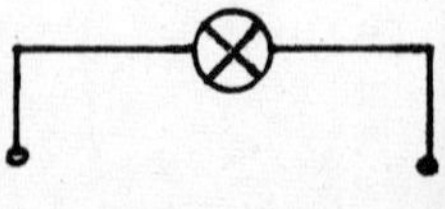

1) The index 'ges' ('gesamt') should be read as 'total'.

```
Rges      [ 1.22569435180E+02 , 1.23898783389E+02 ]
Iges      [ 1.76676472529E+00 , 1.80387549047E+00 ]
U1        [ 5.59964627958E+00 , 5.95059449829E+00 ]
U 2       [ 2.12949405500E+02 , 2.15500353722E+02 ]
```

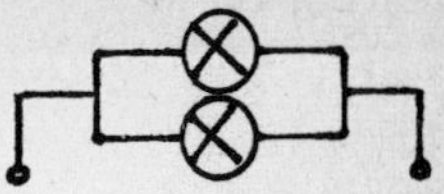

(b)

```
Rges      [ 8.27694351820E+01 , 8.36987833865E+01 ]
Iges      [ 2.61533072694E+00 , 2.67127593073E+00 ]
U1        [ 8.28912121991E+00 , 8.81196065959E+00 ]
U 3       [ 2.10088039338E+02 , 2.12810878782E+02 ]
```

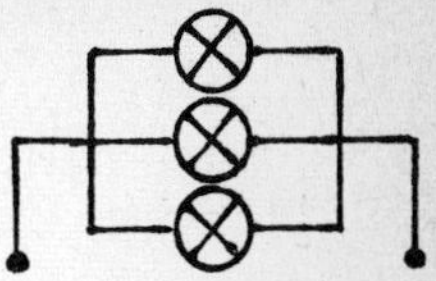

(c)

```
Rges      [ 8.27694351820E+01 , 1.23898783389E+02 ]
Iges      [ 1.76676472529E+00 , 2.67127593073E+00 ]
U1        [ 5.59964627958E+00 , 8.81196065959E+00 ]
U23       [ 2.10088039338E+02 , 2.15500353722E+02 ]
```

(b) or (c)

The operating voltage varies between 210 and 218.3 V .

## Alternating Current Measuring Bridge

The capacity of the unknown condenser C1 and the resistance of the unknown resistor R1 may be found out by using a circuit as shown. The idea is to balance the variable capacity C2 and the variable resistance R2 until the tone in the earphone K (which must be of little resistance) reaches a minimum or vanishes. In this case,

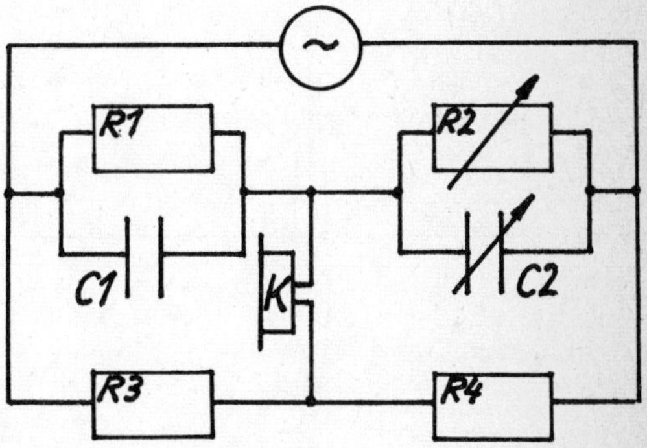


$$C1 = R4 * C2 / R3$$

and

$$R1 = R3 * R2 / R4$$

hold (cf. /G/, S20).

Problem. Given are two resistors of resistance

$$R3 \in [\,9.9,10.1\,]\Omega \quad \text{and } R4 \in [\,6.8,6.9\,]\Omega\ ,$$

according to the producers declaration. Due to u n c e r t a i n t i e s o f p e r c e p t i o n , C2 and R2 are estimated by

$$C2 \in [\,40.2,41.5\,]F \text{ and } R2 \in [\,18.3,19.8\,]\Omega\ .$$

Compute the values of C1 and R1 !

Solution.

```
10    Kanal=7                       ! DRUCKER
20    Adresse=2
30
40    !
50    ! INT INTERVAL R1;R2;R3;R4;C1;C2
60    !
70    ! INT R2:=[18.3,19.8] ! OHM
80    ! INT R3:=[9.9,10.1]  ! OHM
90    ! INT R4:=[6.8,6.9]   ! OHM
100   ! INT C2:=[40.2,41.5] ! FARAD
110   !
120   PRINT "EINGABEDATEN :"
130   PRINT
140   Vorschub=0
150   Genauigkeit=4                 ! 4-STELLIGE AUSGABE
160                                 ! MIT AUSSENRUNDUNG
170   ! INT OUTPUT R2;R3;R4;C2
180   !
190   ! INT C1=R4/R3*C2
200   ! INT R1=R3/R4*R2
210   !
220   PRINT LIN(2);"AUSGABEDATEN :"
230   PRINT
240   Vorschub=1                    ! JEDE AUSGABE EINZELN
250   Genauigkeit=12                ! VOLLE GENAUIGKEIT
260   ! INT OUTPUT C1;R1
270   END
```

Numerical results.

```
EINGABEDATEN :

R2 [1.829E+01,1.981E+01]R3 [9.899E+00,1.011E+01]R4 [6.799E+00,6.901E+00]C2 [4.01
9E+01,4.151E+01]

AUSGABEDATEN :

C1                    [ 2.70653465345E+01 , 2.89242424244E+01 ]
R1                    [ 2.62565217389E+01 , 2.94088235298E+01 ]
```

The condensers capacity is $C1 = (28.0 \pm 1.0)F$, the unknown resistance is $R1 = (27.9 \pm 1.7)\Omega$.

Lens Equation

This example may convince all engineers who "believe" in the classical error estimation method.

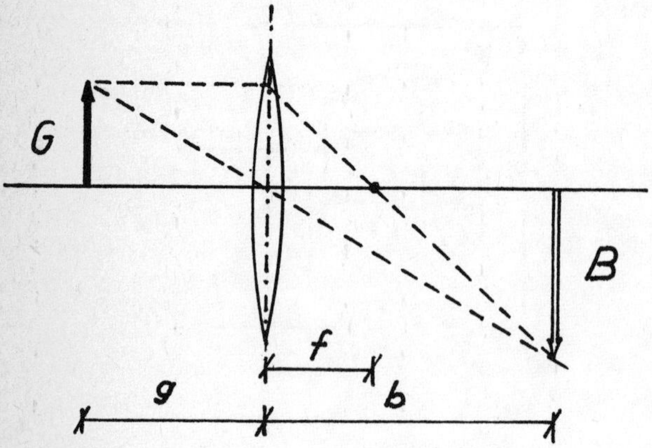


G : object

B : image

f : focal length

g : object distance

b : image distance

Lens eqation for thin lenses :

$1 / f = 1 / b + 1 / g$ .

Problem. Let $f = (20 \pm 1)$cm be the focal length of a thin lens. The image distance b has been metered to $b = (25 \pm 1)$cm. - How large is the distance between the object and the lens ?

Solution. This question is usually handled as follows :

$$g = g(f,b) = 1 / ( 1 / f - 1 / b ),$$
$$g_o = g(f_o, b_o),$$
$$g \doteq g_o \pm \Delta g$$

with
$$\Delta g = ( 1 - f_o / b_o )^{-2} * \Delta f + ( b_o / f_o - 1 )^{-2} * \Delta b .$$

In this case, one has $f_o = 20$cm, $b_o = 25$cm, $\Delta f = \Delta b = 1$. Hence, $g_o = 1 / ( 1/20 - 1/25 ) = 100$, $\Delta g = (1-20/25)^{-2} + (25/20-1)^{-2} = 41$, $g \doteq (100 \pm 41)$cm or

$$g \in [ 59,141 ]\text{cm} .$$

This result is w r o n g !

Calculate instead

$$g \in \lfloor g \rfloor = \underset{\bullet}{1} \oslash ( \underset{\bullet}{1} \oslash \lfloor f \rfloor \ominus \underset{\bullet}{1} \oslash \lfloor b \rfloor )$$

which leads to the ( c o r r e c t ) statement

$$g \in [\ 70.5,168.1\ ]\ \text{cm} .$$

It can easily be seen that the endpoints of this interval ( see the more precise representation in the computer output beneath or the abbreviated version above ) are nearly sharp, i.e. the values can be taken leaving rounding effects aside.

The latter is not true for the result produced with the interval version of the algebraically slightly transformed equation

$$g = ( b * f ) / ( b - f ).$$

This observation helps to explain the need of dependent interval arithmetic or, for practical reasons, of special a l g e b r a i c t r a n s f o r m a t i o n s before evaluating formulas ( see the following examples ).

<u>Numerical results.</u>

```
60    !
70    ! INT INTERVAL Bildweite_b;Brennweite_f;Ggnstandswte_g;Eins
80    !
90    ! INT Bildweite_b:=[24,26]
100   ! INT Brennweite_f:=[19,21]
110   ! INT Eins:=[1]
120   !
130   PRINT "EINGABEDATEN :"
140   PRINT
150   ! INT OUTPUT Bildweite_b;Brennweite_f
160   !
170   ! INT Ggnstandswte_g=Eins/(Eins/Brennweite_f-Eins/Bildweite_b)
180   ! INT Grosses_g=Bildweite_b*Brennweite_f/(Bildweite_b-Brennweite_f)
190   !
200   PRINT LIN(2);"AUSGABEDATEN :"
210   PRINT
220   ! INT OUTPUT Ggnstandswte_g;Grosses_g
230   END

EINGABEDATEN :

Bildweite_b          [ 2.40000000000E+01 , 2.60000000000E+01 ]
Brennweite_f         [ 1.90000000000E+01 , 2.10000000000E+01 ]


AUSGABEDATEN :

Ggnstandswte_g       [ 7.05714285695E+01 , 1.68000000009E+02 ]
Grosses_g            [ 6.51428571425E+01 , 1.82000000002E+02 ]
```

The idea of this problem has been given to the author by his student Berthold SCHOLL.

## <u>Usable Frequency Range of Fiber Optical Waveguides</u>

Look at an optical fiber line of length l with a steplike profile of indices of refraction ( next page ). According to SNELLIUS, one has

$$\sin\alpha / \sin\beta = c_o / c_2 = n_2 / n_o = n_2$$

with $c_o$ velocity of light and $n_o$ aerial index of refraction. The velocity of an axial light pulse is given by $v_2 = c_2 = c_o / n_2$, its time of

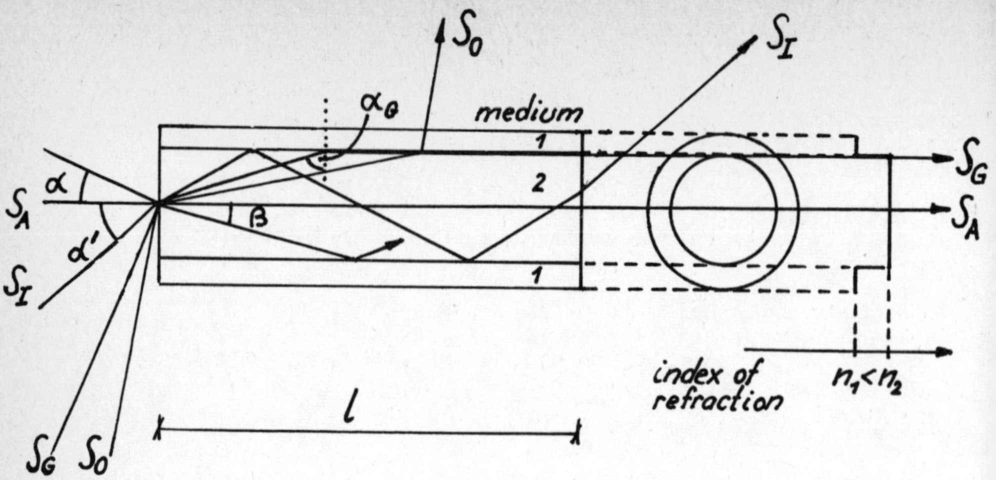


$S_A$ : axial ray, $S_G$ : marginal ray of total reflection, $\alpha_G$ : marginal angle of total reflection, $n_.$ or $c_.$ : index of refraction or velocity of light of medium 0 (air) or 1 or 2 (fiber), respectively.

---

transit is given by $t_A = l / v_2$. The time of transit of a light pulse on the marginal ray $S_G$ is determined by

$$t_G = l / ( v_2 * \sin\alpha_G ) = ( l * n_2 ) / ( v_2 * n_1 ) .$$

Every pulse transmitted at the input side will be received distorted (broadened) on the output side of the line. This is due to the running time difference

$$t_G - t_A > 0.$$

It means that the line cannot be used to transmit pulse sequences of arbitrary choosen frequencies. The usable band width b is to be calculated as follows :

$$b = \frac{1}{2*( t_G - t_A)} = \frac{1}{2*\left( \frac{l*n_2}{v_2*n_1} - \frac{l}{v_2} \right)} = \frac{v_2}{2*l*\left( \frac{n_2}{n_1} - 1 \right)} ,$$

$$b = c_o / ( 2 * l * n_2 * ( n_2 / n_1 - 1 ))$$

(cf. /C/).

Problem. Given a fiber optical waveguide of lenght $l = (220 \pm 0.2)$m. The values of its indices of refraction (steplike index profile assumed) are known with an accuracy of 1 per mil : $n_1 \doteq 1.51$, $n_2 \doteq 1.58$. Estimate the usable band width b by computing

⌊b1⌋ = ⌊co⌋ Ø ( 2̣ ⊛ ⌊l⌋ Ø ⌊n1⌋ ⊛ ⌊n2⌋² - 2̣ ⊛ ⌊l⌋ ⊛ ⌊n2⌋ ) ,

⌊b2⌋ = ⌊co⌋ Ø ( 2̣ ⊛ ⌊l⌋ ⊛ ( ⌊n2⌋² Ø ⌊n1⌋ - ⌊n2⌋ ) ) and

⌊b3⌋ = ⌊co⌋ Ø ( 2̣ ⊛ ⌊l⌋ ⊛ ( ⌊n2⌋ Ø ⌊n1⌋ - 1̣ ) ⊛ ⌊n2⌋ ) .

According to newer mensurations, the velocity $c_o$ of light comes to $(299\ 792\ 456.2 \pm 1.1)$m/s .

Numerical results.

| | | |
|---|---|---|
| L | [ 2.19800000000E+02 | , 2.20200000000E+02 ] |
| N1 | [ 1.50848999999E+00 | , 1.51151000001E+00 ] |
| N2 | [ 1.5784;999999E+00 | , 1.58158000001E+00 ] |
| C0 | [ 2.99792455100E+08 | , 2.99792457300E+08 ] |
| B1 | [ 8.23540866264E+06 | , 1.06834884099E+07 ] |
| B2 | [ 8.53134233801E+06 | , 1.02225602797E+07 ] |
| B3 | [ 8.88314390012E+06 | , 9.76023867943E+06 ] |
| B | [ 8.88314390012E+06 | , 9.76023867943E+06 ] |

There is a guaranteed band width of 8.8MHz. This value is sufficient to perform a telephone communication ( 3.1kHz ) or to transmit music ( 15 kHz ) or a TV program ( 6 MHz ).

Note that the result with the largest l e f t endpoint ( which needs not necessarily to be the result of smallest interval width to fit the practical aspect of the question ) is the most relevant one. The interval

$$[\, 0 \,,\, \llcorner b3 \,]$$

represents all possible frequencies the fiber optic waveguide may be used for.

## Bleeding an Electrical Potential

The configuration shown may be used to transform a given voltage U to a consumer voltage Uv ( at the load resistor Rv ). It is given by

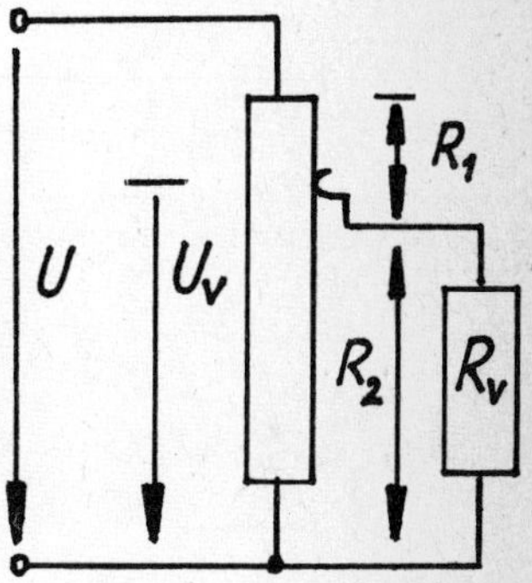


$$U_v = \frac{R_2 * R_v}{R_1 * R_2 + R_1 * R_v + R_2 * R_v} * U ,$$

see /G/, chapter S8.

Problem. What is the range of the consumer potential Uv in a voltage-divider circuit as shown in the figure if U $\in$ [ 210,230 ]V, Rv $\in$ [ 900,910 ]$\Omega$ , R1 $\in$ [ 9.9,10.1 ]$\Omega$ and R2 $\in$ [ 19.8,20.2 ]$\Omega$ ?

Solution. Reduce the formula to the algebraically equivalent form

$$U_v = U / ( R_1 * ( 1 / R_v + 1 / R_2 ) + 1 ).$$

Since each variable does not occur but once in the expression, its interval version will not produce an overestimation.

```
50    ! INT INTERVAL R1;R2;Rv;U;Uv;Eins
60    !
70    ! INT R1:=[9.9,10.1]  ! OHM
80    ! INT R2:=[19.8,20.2] ! OHM
90    ! INT Rv:=[900,910]   ! OHM
100   ! INT U:=[210,230]    ! VOLT
110   ! INT Eins:=[1]
120   !
130   Genauigkeit=4         ! 4-STELLIGE AUSGABE
140                         ! MIT AUSSENRUNDUNG
150   ! INT OUTPUT R1;R2;Rv;U
```

```
160   !
170   ! INT Uv=R2*Rv*U/(R1*R2+R1*Rv+R2*Rv)
180   PRINT LIN(2);"Rv naiv"
190   Vorschub=1                  ! JEDE AUSGABE EINZELN
200   Genauigkeit=12              ! VOLLE GENAUIGKEIT
210   ! INT OUTPUT Uv
220   !
230   ! INT Uv=U/(R1*(Eins/Rv+Eins/R2)+Eins)
240   PRINT LIN(2);"Rv optimiert"
250   ! INT OUTPUT Uv
260   !
270   END
```

Numerical results.

```
R1                    [9.899E+00,1.011E+01]
R2                    [1.979E+01,2.021E+01]
Rv                    [8.999E+02,9.101E+02]
U                     [2.099E+02,2.301E+02]


Rv naiv
Uv                    [ 1.34722875237E+02 , 1.57017635733E+02 ]


Rv optimiert
Uv                    [ 1.38037726326E+02 , 1.53233411791E+02 ]
```

The consumer voltage will vary between 138.0V and 153.3V .

Density Determination of an Unknown Fluid

This example demonstrates the influence of "implicit" constants that can but have not been removed before evaluating a formula. Furthermore, it can be used to teach that sometimes there are error dependent optimal algebraic transformations in absence of a total reduction as could be used in the examples above.

Take a test specimen and handle it as follows :

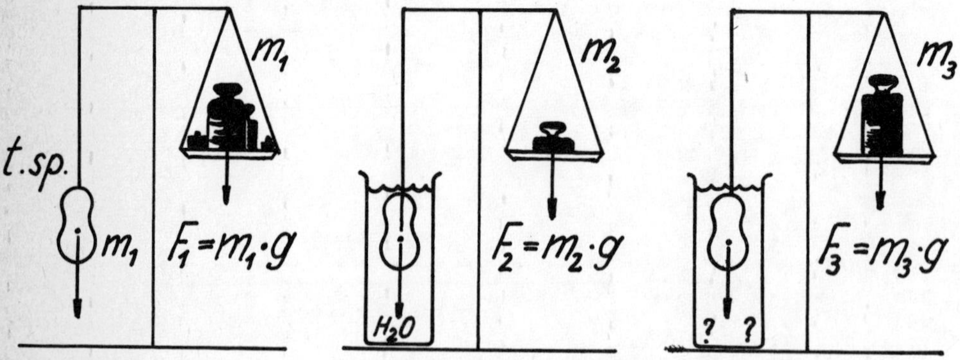


1. Weighing in the air. 2. Weighing in distilled water 3. Weighing in the unknown fluid

Any body of volume V in a fluid of density $\varrho$ meets a lifting force

$$-\varrho * V * g ,$$

where g is the acceleration of the fall. Formally, the following holds :

$$\varrho = \varrho * 1 = \varrho * \frac{V * g}{V * g} = \frac{-\varrho * V * g}{-1 * V * g} .$$

Being aware of the fact that the density of water is $\varrho_{H2O} = 1$, this leads to

$$\varrho_{?} = \frac{-\varrho_{?} * V * g}{-\varrho_{H2O} * V * g} = \frac{F_3 - F_1}{F_2 - F_1} .$$

<u>Problem</u>. Determine the density of an unknown fluid by an amber cube of edge length 1cm using the method explained above. The cube's weight m1 is about 1. Since the density of amber is nearly the same as that of water, $0 < m2 \doteq 0$ will be observed. Let the cube's weight be $m3 \doteq 0.3$ if it is circumcirculated by the unknown fluid. - All masses are given in grams. Assume at first an accuracy of $\pm$ 0.001, then change the error successively for each mass from $\pm$ 0.001 to $\pm$ 0.005. Compute for all four cases four different values :

$$\lfloor \varrho 1 \rfloor = ( \lfloor F3 \rfloor \ominus \lfloor F1 \rfloor ) \oslash ( \lfloor F2 \rfloor \ominus \lfloor F1 \rfloor ) ,$$

$$\lfloor \varrho 2 \rfloor = ( \lfloor m1 \rfloor \ominus \lfloor m3 \rfloor ) \oslash ( \lfloor m1 \rfloor \ominus \lfloor m2 \rfloor ) ,$$

$$\lfloor \varrho 3 \rfloor = \underset{\bullet}{1} \oplus ( \lfloor m2 \rfloor \ominus \lfloor m3 \rfloor ) \oslash ( \lfloor m1 \rfloor \ominus \lfloor m2 \rfloor ) ,$$

$$\lfloor \varrho 4 \rfloor = \underset{\bullet}{1} \oslash ( \underset{\bullet}{1} \oplus ( \lfloor m3 \rfloor \ominus \lfloor m2 \rfloor ) \oslash ( \lfloor m1 \rfloor - \lfloor m3 \rfloor ) ) .$$

<u>Solution</u>.

```
30    !
40    ! INT INTERVAL F<1:3,1:4>;M<1:3,1:4>;Rho<1:4,1:4>
50    ! INT INTERVAL Mwert<3>;Fehler;Eins;G
60    !
70    ! INT Mwert<1>:=[.999,1.001]          ! MASSEN GENAU
80    ! INT Mwert<2>:=[0,.001]
90    ! INT Mwert<3>:=[.299,.301]
100   ! INT Fehler:=[-.004,.004]            ! FEHLERZUSCHALG
110   ! INT G:=[9.8063,9.8151]              ! FALLBESCHLEUNIGUNG
120   ! INT Eins:=[1]                       ! PUNKTINTERVALL
130   !
140   FOR I=1 TO 4
150      FOR J=1 TO 3
160         ! INT M<J,I>:=Mwert<J>          ! MASSEN DEFINIEREN
170      NEXT J
180      IF I>1 THEN
190         K=I-1
200          ! INT M<K,I>=M<K,I>+Fehler     ! FEHLER ANBRINGEN
210         IF K=2 THEN M(2,I,1)=0          ! MASSE NICHT NEGATIV
220      END IF
230   NEXT I
240   FOR I=1 TO 4
250      FOR J=1 TO 3
260         ! INT F<J,I>=M<J,I>*G           ! KRAEFTE BERECHNEN
270      NEXT J
280   NEXT I
290   FOR I=1 TO 4                          ! SPEZ. GEWICHT
300      ! INT  Rho<1,I>=(F<3,I>-F<1,I>)/(F<2,I>-F<1,I>)
310      ! INT  Rho<2,I>=(M<1,I>-M<3,I>)/(M<1,I>-M<2,I>)
320      ! INT  Rho<3,I>=Eins+(M<2,I>-M<3,I>)/(M<1,I>-M<2,I>)
330      ! INT  Rho<4,I>=Eins/(Eins+(M<3,I>-M<2,I>)/(M<1,I>-M<3,I>))
340   NEXT I
350   !
```

Numerical results.

| | |
|---|---|
| M1 | [ 9.989E-01, 1.002E+00] |
| M2 | [-1.000E-99, 1.001E-03] |
| M3 | [ 2.989E-01, 3.011E-01] |
| Rho1 | [ 6.963E-01, 7.044E-01] |
| Rho2 | [ 6.972E-01, 7.035E-01] |
| Rho3 | [ 6.983E-01, 7.024E-01] |
| Rho4 | [ 6.986E-01, 7.021E-01] |

——— worst of all results
········ worst result leaving Rho1 aside

| | | | |
|---|---|---|---|
| M1 | [ 9.949E-01, 1.006E+00] | [ 9.989E-01, 1.002E+00] | [ 9.989E-01, 1.002E+00] |
| M2 | [-1.000E-99, 1.001E-03] | [-1.000E-99, 5.001E-03] | [-1.000E-99, 1.001E-03] |
| M3 | [ 2.989E-01, 3.011E-01] | [ 2.989E-01, 3.011E-01] | [ 2.949E-01, 3.051E-01] |
| Rho1 | [ 6.896E-01, 7.113E-01] | [ 6.963E-01, 7.072E-01] | [ 6.923E-01, 7.084E-01] |
| Rho2 | [ 6.904E-01, 7.104E-01] | [ 6.972E-01, 7.063E-01] | [ 6.932E-01, 7.075E-01] |
| Rho3 | [ 6.971E-01, 7.036E-01] | [ 6.971E-01, 7.064E-01] | [ 6.943E-01, 7.064E-01] |
| Rho4 | [ 6.974E-01, 7.033E-01] | [ 6.986E-01, 7.049E-01] | [ 6.946E-01, 7.061E-01] |

It is obvious, that the formula for $\lfloor \varrho 1 \rfloor$ is bad in all cases since the superfluous constant g has not been removed in time . More surprising is, that $\lfloor \varrho 4 \rfloor$ produces ( although using 5 instead of 4 or 3 arithmetic operations ! ) the best result for all cases. It does so even if the error of m3 is the largest one. This is a contradiction to the idea that the number of occurrences of a large width interval variable in a formula should be minimized in any case.
Remark : For the choice of the interval $\lfloor g \rfloor$ ( acceleration of the fall ) see the last example. - Known fluid densities next to the results above are those of petro ether ( 0.67kg/dm$^3$ ), benzine or hydrocyanic acid ( 0.7 kg/dm$^3$ each ), ether ( 0.73kg/dm$^3$ ) and alcohol or acetone ( 0.79kg/dm$^3$ each ). If the fluid is no emulsion and the experimenter is still alive, all results indicate benzine.

## Hydrostatic Pressure in Open Reservoirs

The last example may help to illustrate why there is a need for functions of type $\mathbb{R} \rightarrow I(\mathbb{R})$ or even $I(\mathbb{R}) \rightarrow I(\mathbb{R})$.

Consider a fluid of known density $\varrho$ in an open reservoir. Let the bathometer be normed as shown in the figure. Let $p_B$ be the hydrostatic pressure at z = 0. Then follows for the hydrostatic pressure

$$p(z) = p_B + g * \varrho * z .$$

If $p_B \in \lfloor ao \rfloor$ and $g * \varrho \in \lfloor a1 \rfloor$, one has

$$p(z) \in \lfloor a0 \rfloor \oplus \lfloor a1 \rfloor \circledast \dot{z}$$

with an s t r a i g h t l i n e i n t e r v a l on the right hand side.

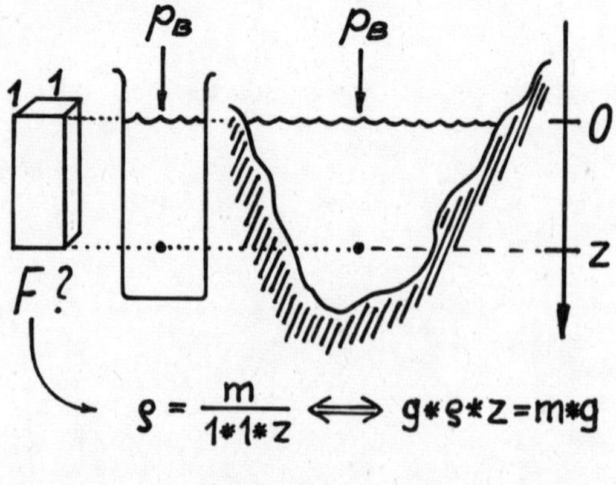


Problem. Plot a diagram / calculate a table for a diver to show the underwater pressure for bar ometric air pressures between 930mbar and 1070 mbar (HPa) and depths of water between 0m and 100m. - Take into account, that the density of natural water (depending upon its salt content ) varies between 0.99kg/dm³ and 1.03kg/dm³ and that the acceleration g of the fall differs in its value depending on the terrestrial latitude :

| Latitude | 0° | 20° | D 40° | D 45° | D 50° | 55° | 60° | 90° |
|---|---|---|---|---|---|---|---|---|
| $m/s^2$ ±1E-4 | 9.7805 | 9.7865 | 9.8018 | 9.8063 | 9.8108 | 9.8151 | 9.8192 | 9.8322 |

D : Germany

( cf. /GE/ ). - For practical reasons, use steps of 20 for the barometric air pressure.

Solution.

```
10    ! INT INTERVAL P;Pb;G;Rho;Z;Anstieg;Zuschlag  ! DEKLARATION
20    ! INT G:=[9.7804,9.8323]                      ! FALLBESCHLEUNIGUNG
30    ! INT Rho:=[.99,1.03]                         ! DICHTE H20
40    ! INT Anstieg=G*Rho                           ! GERADENANSTIEG
50    !
        . . .
420     !
430     ! INT Zuschlag:=Anstieg                     ! GERADEN ZEICHNEN
440     Zuschlag(1)=Zuschlag(1)*100
450     Zuschlag(2)=Zuschlag(2)*100
460     FOR X=930 TO 1050 STEP 20
470        ! INT Pb:=[X,X+20]
480        ! INT P=Pb+Zuschlag
490        MOVE Pb(1),0                             ! LINKE ECKE
500        LINE TYPE 1
510        DRAW P(1),-100
520        MOVE P(2),-100                           ! RECHTE ECKE
530        LINE TYPE 4
540        DRAW Pb(2),0
550     NEXT X
560     DUMP GRAPHICS #Kanal,Adresse
570     END
```

( All parts of the program that do not use interval arithmetic have been suppressed. )

Graphic result. The diagram is given on the next page.

A pearl-fisher will not reach a deepness of more than 35m. Assume a barometric air pressure of 1000mbar. The diagram says that he will suffer a water pressure of at most between 1320mbar and 1370mbar.

A man who uses a diving dress may reach 90m to 100m. According to the diagram, he will find a water pressure of between 1860mbar and 2030mbar at that depth.

It might be interesting to plot one straight line interval for PICARD and WALSH that reached in 1960 with their bathyscaph "Trieste" the depth of 10912m. Since the air pressure of the two days experiment may not be available, take [ 930,1070 ]mbar, the interval of all possible natural values. Otherwise take the interval observed during the 22nd and 23rd of January 1960 by the experimenters.

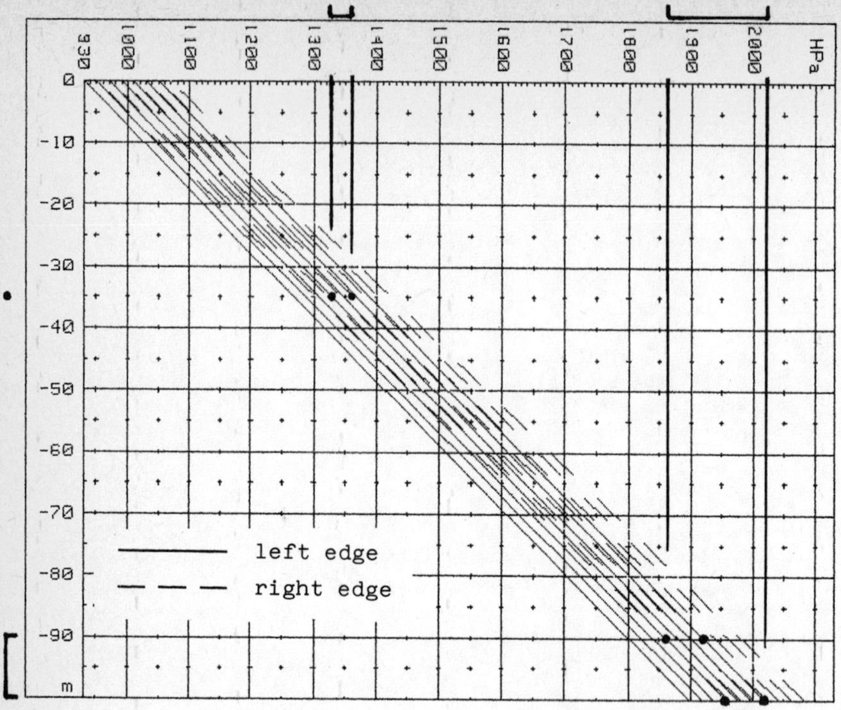


For more examples see /T/.

Literature


/C/ Conrad CHRISTIANI
Elektroniklabor
Christiani-Verlag
1984

/G/ Kurt GIECK
Technische Formelsammlung
Gieck-Verlag, Heilbronn
25. Aufl.(deutsch), 1977

/GE/ Walter GELLERT (Hrsg.) et al.
Kleine Enzyklopädie Natur
VEB Bibliographisches Institut
Leipzig 1966

/L/ Helmut LINDNER
Physikalische Aufgaben
Verlag Friedrich Vieweg
Braunschweig
8. Aufl. 1966

/T/ Peter THIELER
Technisches Rechnen mit
Intervallen
To be submitted.


Remark : All the standard literature on interval mathematics needed in this paper has been ommitted for the sake of shortness.

# GENERALIZED THEORY AND SOME SPECIALIZATIONS OF THE REGION CONTRACTION ALGORITHM I - BALL OPERATION

You Zhaoyong, Xu Zongben, Liu Kunkun

Mathematics Department of
Xi'an Jiaotong University
Xi'an
The People's Republic of China

Abstract.

We describe a new algorithm named Region Contraction Algorithm for solving certain nonlinear equations, and establish the convergence of the algorithm and give an error estimation. It is shown that this general theory includes all of present existing ball iterations as special cases.

To find a zero of a quasi-strongly monotone mapping, which arises often from the field of differential equations, variational calculus and optimization etc., the authors [2] recently proposed a new algorithm called Region Contraction Algorithm (abbreviated RCA henceforth) in real Hilbert spaces. Stemming from T.E. Williamson's geometric estimation for fixed points of contractive mappings [3], the algorithm establishes a convergent iterative process which keeps well defined and automatically covers the errors by constructing a sequence of closed balls containing the zero set. Later on, proceeding in a completely different view from the authors, Wu Yujiang and Wang Deren [4] rewrited our algorithm in the language of interval analysis, and also suggested a new globally convergent scheme in the case that F is strongly monotone. It showed the authors that the RCA is almost Nickel's Ball Newton Method [1] (abbreviated BNM henceforth) except for the difference of the class of mappings to which it applies.

In this paper we develop a more general algorithm called stationary region contracting algorithm (abbreviated SRCA) with RCA, BNM and some other methods as its specializations.

In Section 1 we present the algorithm and give some basic properties in Section 2. In Section 3 we prove convergence of the algorithm and discuss some specializations in the last section.

In what follows, we always let H be a real Hilbert space with inner product (.,.), and use B(x,r) to denote the closed ball with center x and radius r.

## 1. Algorithm.

Let D be a subset of H, $B(b,d) \subset D$ a given closed ball with $d > 0$, and $F: D \subset H \to H$ a given nonlinear mapping. We want to find a zero of the mapping F in the ball B(b,d). Let us suppose that there exists a nonlinear mapping $g: B(b,d) \to H$ and a nonnegative functional $r: B(b,d) \to R^+$ such that for all $x \in B(b,d)$

$$r(x) \leqq \|g(x)\| \tag{1.1}$$

and

$$N(F) \subset Gx, \tag{1.2}$$

where N(F) is the set of zeros of F in B(b,d) and Gx is defined as

$$Gx = B(x - g(x),\ r(x)).$$

For any two closed balls B' and B", let <B' ∩ B"> stand for the minimum-volume closed ball which contains their intersection if $B' \cap B'' \neq \emptyset$, and $\langle B' \cap B''\rangle = \emptyset$ if $B' \cap B'' = \emptyset$.

We develop our general algorithm SRCA as follows:

I. Initial Step
Set $B_0 = B(x_0, r_0) = B(b,d)$.

II. Continuation Step
Suppose that $B(x_k, r_k)$ has been constructed; we then continue to construct the next ball $B_{k+1}$ in the following way:

II.1. Starting Step. If $r_k = 0$, then stop the algorithm at (*) when $Fx_k \neq 0$, otherwise (**) when $Fx_k = 0$. If $r_k \neq 0$, then calculate $Gx_k$.

II.2. Contraction Step. Stop the algorithm at (*) if $Gx_k \cap B_0 = \emptyset$, otherwise set $\bar{B}_{k+1} = B(\bar{x}_{k+1}, \bar{r}_{k+1}) = \langle Gx_k \cap B_k \rangle$.

II.3. Modification Step. Stop the algorithm at (*) if $\bar{B}_{k+1} \cap B_0 = \emptyset$, otherwise set

$$B_{k+1} = B(x_{k+1}, r_{k+1}) = \begin{cases} \bar{B}_{k+1} & \text{if } \bar{x}_{k+1} \in B_0, \\ \langle \bar{B}_{k+1} \cap B_0 \rangle & \text{if } \bar{x}_{k+1} \notin B_0 . \end{cases}$$

III. Return to II. with k := k+1.

A continuation step from $B_k$ to $B_{k+1}$ for SRCA is shown in the following figures (where $H = R^2$).

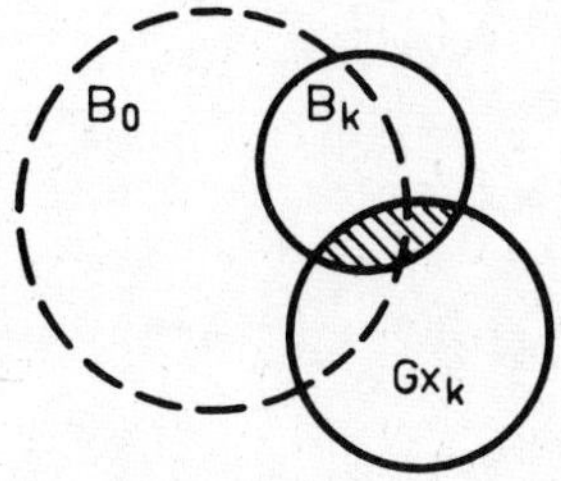


II.1.

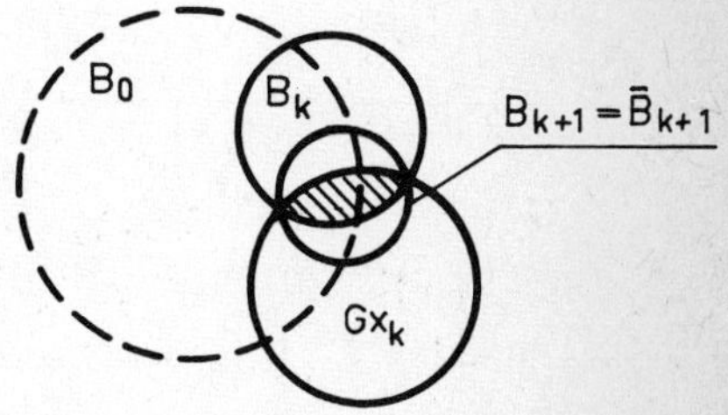


II.2. and II.3. $\bar{x}_{k+1} \in B_0$

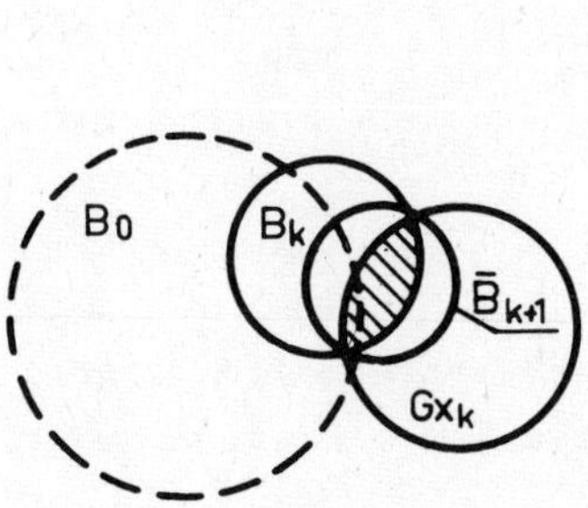


II.1. and II.2.

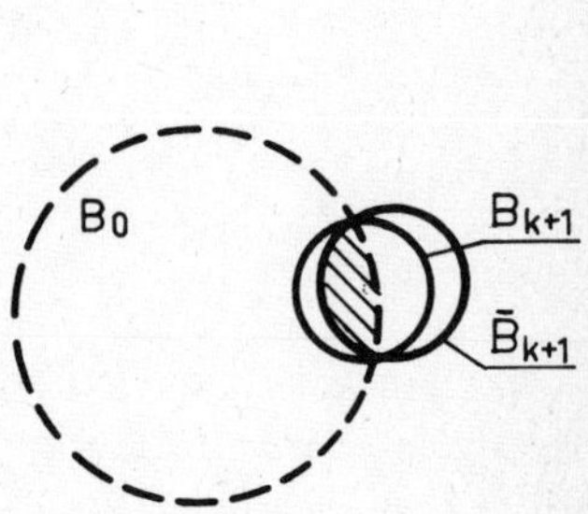


II.2. and II.3. $\bar{x}_{k+1} \notin B_0$

Remark. (1) It is easily seen (see Theorem 1 in next section) that the SRCA generates successively a sequence of closed balls which contain N(F), whose centers are located in $B_0$, and whose radii decrease step by step. Thus, the algorithm is always well defined, the error of computation is automatically covered, and we may expect to find N(F).

(2) If $B' = B(c_1,s_1)$ and $B'' = B(c_2,s_2)$ are any two closed balls, and $s_1 \geq s_2 > 0$, then $\langle B' \cap B'' \rangle$ can be represented by the following formula (see Lemma 2 in [3] and [1]):

$$\langle B' \cap B'' \rangle = \begin{cases} \emptyset & \text{if } \|c_1 - c_2\| > s_1 + s_2, \\ B'' & \text{if } \|c_1 - c_2\| \leq (s_1^2 - s_2^2)^{\frac{1}{2}}, \\ B(c,s) & \text{otherwise,} \end{cases} \tag{1.3}$$

where

$$\begin{aligned} c &= w_1 c_1 + (1 - w_1) c_2 = (1 - w_2) c_1 + w_2 c_2 \in B' \cap B'', \\ s &= (s_2^2 - w_1^2 \|c_1 - c_2\|^2)^{\frac{1}{2}} = (s_1^2 - w_2^2 \|c_1 - c_2\|^2)^{\frac{1}{2}}, \\ w_1 &= \tfrac{1}{2}(1 - (s_1^2 - s_2^2) / \|c_1 - c_2\|^2) = 1 - w_2. \end{aligned} \tag{1.4}$$

Particularly, if $\langle B' \cap B'' \rangle = B(\xi,\eta) \neq \emptyset$, then $\xi \in B' \cap B''$ and $\eta \leq \min(s_1,s_2)$.

(3) A natural idea is that one always takes $B_{k+1} = \langle \bar{B}_{k+1} \cap B_0 \rangle$ in the Modification Step of II.3., but a simple analysis shows that this modification affects only weakly the convergence of the algorithm, and hence, there is no need to do so for saving time of computation.

## 2. Basic Properties.

Let $\{B_k\}$ be the sequence of closed balls generated by the algorithm, and for convenience, we define $B_j = B_k$ for all $j > k$ if the algorithm stops at (**) in the (k+1)-th step.

Theorem 1.

The following statements are all valid:

(1) $x_k \in B_o$ for each k .

(2) $N(F) \subset B_k$ for each k .

(3) If the algorithm stops in the $(k_o+1)$-th step, then either $N(F) = \emptyset$ if (*) appears or $N(F) = \{x^*\}$ if (**) appears.

(4) If $N(F) \neq \emptyset$ and $r^* = \lim_{(k)} r_k = 0$, then $x_k$ converges to $x^*$, the unique element of $N(F)$, and

$$\|x_k - x^*\| \leq r_k \qquad \forall\, k \geqq 0.$$

Proof. (1) and (2): It is obvious for k = 0. Suppose that the conclusions hold for some k > 0 and $B_{k+1}$ can be produced, then by (1.2) $N(F) \subset Gx_k \cap B_k \subset \langle Gx_k \cap B_k\rangle = \overline{B}_{k+1}$. If $\overline{x}_{k+1} \in B_o$, then $B_{k+1} = \overline{B}_{k+1}$ and $x_{k+1} = \overline{x}_{k+1}$ by the definition of the algorithm; otherwise, $B_{k+1} = \langle\overline{B}_{k+1} \cap B_o\rangle$, and hence $N(F) \subset \overline{B}_{k+1} \cap B_o \subset B_{k+1}$ and $x_{k+1} \in \overline{B}_{k+1} \cap B_o \subset B_o$. Therefore, the conclusion also holds for the case k+1. By induction, (1) and (2) are valid.

(3): If the algorithm stops at (*), then $N(F) = \emptyset$ because it comes about if and only if one of the following cases occurs: either $B_{k_o} = \{x_{k_o}\} \subset B_o$ and $Fx_{k_o} \neq 0$, or $Gx_{k_o} \cap B_o = \emptyset$ or $\overline{B}_{k_o+1} \cap B_o = \emptyset$. If the algorithm stops at (**), then $B_{k_o} = \{x_{k_o}\}$ and $Fx_{k_o} = 0$, i.e., $N(F) = \{x^*\}$ with $x^* = x_{k_o}$ .

(4): If $N(F) \neq \emptyset$, the algorithm is never stopped at (*), so $\{r_k\}$ is infinite. Therefore, the result is direct consequence of (2) and (3).

Theorem 2.

The sequence $\{r_k\}$ decreases in the following way

$$\overline{r}_{k+1} \leq \begin{cases} r_k/\, 2^{\frac{1}{2}} & \text{if } \|g(x_k)\|^2 + r(x_k)^2 \leq r_k^2, \\ (1 - r_k^2/(4\|g(x_k)\|^2))^{\frac{1}{2}} r_k & \text{otherwise} \end{cases} \tag{2.1}$$

and

$$r_{k+1} \leq \begin{cases} \overline{r}_{k+1} & \text{if } \overline{x}_{k+1} \in B_o, \\ (1 - \overline{r}_{k+1}^2/(4d^2))^{\frac{1}{2}}\, \overline{r}_{k+1} & \text{if } \overline{x}_{k+1} \overline{\in} B_o. \end{cases} \tag{2.2}$$

Proof. If $\|g(x_k)\|^2 + r(x_k)^2 \leqq r_k^2$, then $\bar{B}_{k+1} = \langle Gx_k \cap B_k \rangle = Gx_k$ by (1.3) and hence (1.1) yields that

$$2\bar{r}_{k+1}^2 = 2r(x_k)^2 \leqq \|g(x_k)\|^2 + r(x_k)^2 \leqq r_k^2,$$

that is, $\bar{r}_{k+1} \leqq r_k / 2^{\frac{1}{2}}$. If $\|g(x_k)\|^2 + r(x_k)^2 > r_k^2$, then it follows from (1.1) that $\|g(x_k)\|^2 \geqq |r_k^2 - r(x_k)^2|$, so (1.1) and (1.3)-(1.4) give that

$$\bar{r}_{k+1} = [r_k^2 - (\|g(x_k)\|^2 + r_k^2 - r(x_k)^2)^2 / (2\,\|g(x_k)\|)^2]^{\frac{1}{2}}$$

$$\leqq [r_k^2 - (r_k^2/(2\|g(x_k)\|^2)]^{\frac{1}{2}} = r_k(1 - r_k^2/(4\|g(x_k)\|^2))^{\frac{1}{2}} \;;$$

therefore, (2.1) follows.

Furthermore, if $\bar{x}_{k+1} \in B_0$, then $r_{k+1} = \bar{r}_{k+1}$ because $B_{k+1} = \bar{B}_{k+1}$. If $\bar{x}_{k+1} \bar{\in} B_0$, then $\|\bar{x}_{k+1} - b\| > d$ and $B_{k+1} = \langle \bar{B}_{k+1} \cap B_0 \rangle$, and hence one gets

$$r_{k+1} \leqq (1 - \bar{r}_{k+1}^2/(4d^2))^{\frac{1}{2}}\, \bar{r}_{k+1}$$

from Lemma 3 in [2]. Thus, the proof of (2.2) is completed.

A simple consequence of Theorem 2 is that $r^* = \lim_{(k)} r_k$ always exists if the algorithm is never stopped at (*).

Theorem 3.
For each pair $m \geqq n > 0$, $\{B_k\}$ satisfies that

$$r_m^2 \leqq r_n^2 - \sum_{k=n}^{m-1} \|x_{k+1} - x_k\|^2 .$$

Proof. Obviously, it is only necessary to prove the following inequality

$$\|x_{k+1} - x_k\|^2 \leqq r_k^2 - r_{k+1}^2 \tag{2.3}$$

for each $k \geqq 0$ in the case that $B_{k+1}$ can be produced. We complete the proof considering two cases:

(1) $\bar{x}_{k+1} \in B_0$, then $B_{k+1} = \bar{B}_{k+1} = \langle Gx_k \cap B_k \rangle$. As done in the proof of Theorem 2, $\|g(x_k)\|^2 + r(x_k)^2 \leqq r_k^2$ implies that $B_{k+1} = \langle Gx_k \cap B_k \rangle = Gx_k$ and hence

$$\|x_{k+1} - x_k\|^2 = \|g(x_k)\|^2 \leqq r_k^2 - r(x_k)^2 = r_k^2 - r_{k+1}^2. \tag{2.4}$$

Similarly, from $\|g(x_k)\|^2 + r(x_k)^2 > r_k^2$, we get that $\|g(x_k)\|^2 \geqq$ $\geqq |r_k^2 - r(x_k)^2|$, so (1.3)-(1.4) gives directly that

$$\|x_{k+1} - x_k\|^2 = r_k^2 - r_{k+1}^2 .$$

Therefore, (2.3) holds in this case.

(2) $\bar{x}_{k+1} \overline{\in} B_0$, then $B_{k+1} = \langle \bar{B}_{k+1} \cap B_0 \rangle$. Since $\bar{r}_{k+1} \leqq r_k \leqq r_0 = d$ by Theorem 2, it follows that

$$\|\bar{x}_{k+1} - b\| > d \geqq (d^2 - \bar{r}_{k+1}^2)^{\frac{1}{2}}$$

and hence we conclude from (1.3)-(1.4) that $x_{k+1} = w_k b + (1-w_k)\bar{x}_{k+1}$ and

$$r_{k+1}^2 = \bar{r}_{k+1}^2 - w_k^2 \|\bar{x}_{k+1} - b\|^2 , \tag{2.5}$$

where

$$w_k = \tfrac{1}{2}(1 - (d^2 - \bar{r}_{k+1}^2) / \|\bar{x}_{k+1} - b\|^2). \tag{2.6}$$

Noting the following identity

$$\begin{aligned} \|x_{k+1} - x_k\|^2 &= \|w_k(b - x_k) + (1-w_k)(\bar{x}_{k+1} - x_k)\|^2 \\ &= w_k\|b - x_k\|^2 + (1-w_k)\|\bar{x}_{k+1} - x_k\|^2 - w_k(1-w_k)\|\bar{x}_{k+1} - b\|^2 \end{aligned}$$

and the fact that (2.4) indicates that $\|\bar{x}_{k+1} - x_k\|^2 \leqq r_k^2 - \bar{r}_{k+1}^2$, we obtain

$$\begin{aligned} \|x_{k+1} - x_k\|^2 &\leqq w_k d^2 + (1-w_k)(r_k^2 - \bar{r}_{k+1}^2) - w_k(1-w_k)\|\bar{x}_{k+1} - b\|^2 \\ &= (1-w_k)(r_k^2 - \bar{r}_{k+1}^2) + w_k\bar{r}_{k+1}^2 + w_k((d^2 - \bar{r}_{k+1}^2)/\|\bar{x}_{k+1} - b\|^2 \\ &\quad -1 + w_k)\ \|\bar{x}_{k+1} - b\|^2 \\ &= (1-w_k)r_k^2 + (2w_k - 1)\bar{r}_{k+1}^2 - w_k^2\ \|\bar{x}_{k+1} - b\|^2. \end{aligned} \tag{2.7}$$

Combining (2.5) with (2.7), we therefore know a sufficient condition for (2.3) holds is the following

$$(1-w_k)r_k^2 + (2w_k - 1)\bar{r}_{k+1}^2 - w_k^2 \|\bar{x}_{k+1}-b\|^2 \leqq r_k^2 - \bar{r}_{k+1}^2 + w_k^2 \|\bar{x}_{k+1} - b\|^2$$

or equivalently,

$$2\bar{r}_{k+1}^2 - r_k^2 \leqq 2w_k \|\bar{x}_{k+1} - b\|^2. \qquad (2.8)$$

But from (2.6),

$$2w_k \|\bar{x}_{k+1} - b\|^2 = \|\bar{x}_{k+1} - b\|^2 - d^2 + \bar{r}_{k+1}^2$$

so (2.8) is equivalent to

$$\bar{r}_{k+1}^2 + d^2 \leqq r_k^2 + \|\bar{x}_{k+1} - b\|^2 .$$

Consequently, the validity of (2.3) immediately follows from $\bar{r}_{k+1} \leqq r_k$ and $d \leqq \|\bar{x}_{k+1} - b\|$.

Remark. Theorems 1-3 generalize Theorems 1-3 of [2]; for a nonexpansive mapping T.E. Williamson, Jr. has proved a similar estimation as in Theorem 3 (see Theorem 6 in [3]).

Using the parameter $\lambda \in [0,1]$ defined by

$$\lambda = \operatorname{Sup}\{\, r(x)/\|g(x)\| \;;\; x \in B_0 \text{ and } g(x) \neq 0 \,\}$$

we characterice now the algorithm in another way.

Theorem 4.
The sequence $\{r_k\}$ decreases in the following form

$$r_k \leqq \lambda r_{k-1} \leqq \lambda^k d.$$

Proof. For any possible $k \geq 0$, if $\|g(x_k)\|^2 + r(x_k)^2 \leqq r_k^2$, then $\|g(x_k)\| \leqq r_k$, and hence $\bar{r}_{k+1} \leqq r(x_k) = (r(x_k)/\|g(x_k)\|)\,\|g(x_k)\| \leqq \lambda r_k$; if $\|g(x_k)\|^2 + r(x_k)^2 > r_k^2$, then $\|g(x_k)\| \geqq |r_k^2 - r(x_k)^2|^{\frac{1}{2}}$ and by (1.3)-(1.4)

$$\bar{r}_{k+1}^2 = r_k^2 - [(\|g(x_k)\|^2 + r_k^2 - r(x_k)^2)/(2\|g(x_k)\|)]^2. \qquad (2.9)$$

Hence, from the simple inequality $2ab \leqq a+b$, it follows that

$$2r_k(\|g(x_k)\|^2 - r(x_k)^2)^{\frac{1}{2}} \leqq r_k^2 + (\|g(x_k)\|^2 - r(x_k)^2);$$

therefore we conclude from (2.9) that

$$\bar{r}_{k+1}^2 \leqq r_k^2 - [2r_k(\|g(x_k)\|^2 - r(x_k)^2)^{\frac{1}{2}} / (2\|g(x_k)\|)]^2$$

$$= (r(x_k)/\|g(x_k)\|)^2 r_k^2 \leqq \lambda^2 r_k^2,$$

i.e., $\bar{r}_{k+1} \leqq \lambda r_k$. Thus, the conclusion immediately follows from the fact $r_{k+1} \leqq \bar{r}_{k+1}$.

## 3. Convergence

We now establish the convergence of the SRCA. In the first result, a generalization to Theorem 2 of [2], we only presuppose the boundedness of g.

Theorem 5 (Bounded Convergence).
If g is bounded on $B_0$ and $N(F) \neq \emptyset$, then $x_k$ converges to $x^*$, the unique element of $N(F)$, in the following way

$$\|x_k - x^*\| \leqq r_k \to 0 \text{ as } k \to \infty.$$

Proof. By (4) of the Theorem 1, it is sufficient to show $r^* = \lim r_k = 0$. We assume the contrary, namely that $r^* > 0$ and the algorithm never stops. Since g is bounded on $B_0$, there is a constant $M > r^*$ such that

$$\operatorname{Sup}\{\|g(x)\|\,;\ x \in B_0\} \leqq \tfrac{1}{2}M.$$

It follows from $x_k \in B_0$ that $2\|g(x_k)\| \leqq M$ for all $k \geqq 0$. Thus, we have for all $k \geqq 0$ that

$$r_k^2/(4\|g(x)\|^2) \geqq r^{*2}/M^2$$

and hence from Theorem 2 that

$$r_{k+1} \leqq \bar{r}_{k+1} \leqq qr_k, \tag{3.1}$$

where

$$q = \text{Max}\{ 2^{-\frac{1}{2}} , (1 - r^*/M^2)^{\frac{1}{2}} \} < 1.$$

Taking the limit as $k \to \infty$, we get the obvious contradiction: $r^* \leqq qr^* < r^*$ which shows that $r^*$ must be zero.

Generally speaking, some additional conditions are necessary for guaranteeing the sequence $\{B_k\}$ shrink to $N(F)$. We below consider a important particular case in which g and r are of the following forms

$$g(x) = u(x)P(Fx) \tag{3.2}$$

$$r(x) = v(x) \| P(Fx) \|, \tag{3.3}$$

where u and v are nonnegative functionals on $B_0$, and $P: H \to H$ is a mapping with 0 as its unique zero.

Recall that a mapping $T: C \subset H \to H$ is said to be closed if its graph $\{(x,Tx) \in H \times H;\ x \in C\}$ is closed in the product space $H \times H$.

<u>Theorem 6</u> (Global convergence).
If $\lambda < 1$, then $\{B_k\}$ shrinks as $k \to \infty$ to a singleton containing $N(F)$ provided the algorithm never stops at (*). In addition, if the composed mapping PF restricted to $B_0$ is closed (especially, continuous), and

$$\varepsilon = \text{Inf}\{ v(x);\ x \in B_0 \text{ and } u(x)P(Fx) \neq 0\} > 0 \tag{3.4}$$

then

(1) $N(F) = \emptyset$ iff the algorithm terminates at (*);

(2) $N(F) \neq \emptyset$ iff the algorithm never terminates at (*), and in this case $x_k$ converges to $N(F) = \{x^*\}$ with the following estimation

$$\| x_k - x^* \| \leqq r_k \leqq \lambda^k d.$$

Proof. If $\lambda < 1$ and the algorithm never stops at $(*)$, then $\{B_k\}$ is infinite, and $r_k \to 0$ as $k \to \infty$ by Theorem 4. Also, since $\|x_{k+1} - x_k\| \leq r_k \leq d\lambda^k$ by (2.3) and Theorem 4, $\{x_k\}$ is a Cauchy sequence so that it is convergent, and hence the first part follows.

For the last part, by the definition of the algorithm and $\{B_k\}$, we only need to prove the sufficiency of (2). Suppose that the algorithm never stops at $(*)$, then $\{B_k\}$ must be infinite. If the algorithm stops at $(**)$, the conclusion follows directly from (4) of Theorem 1, so it remains to discuss the case that the algorithm never stops.

For any $k \geq 0$, $\bar{B}_{k+1} = \langle Gx_k \cap B_k \rangle \neq \emptyset$ implies that $\|g(x_k)\| \leq r(x_k)+r_k$. It follows from (3.2)-(3.3) that

$$r(x_k) = v(x_k)\|P(Fx_k)\| = \|u(x_k)P(Fx_k)\| \, v(x_k)/u(x_k)$$

$$= \|g(x_k)v(x_k)/u(x_k)\| \leq (r(x_k)+r_k)\lambda$$

and hence $r(x_k) \leq \lambda(1-\lambda)^{-1} r_k$. Thus, we conclude from (3.4) that

$$\|P(Fx_k)\| = r(x_k)/v(x_k) \leq \lambda r_k/(1-\lambda)\varepsilon$$

which shows $P(Fx_k) \to 0$ as $k \to \infty$. On the other hand, by the first part of the Theorem, there exists an $x^* \in B_0$ such that $x_k \to x^*$ as $k \to \infty$ and we have that $P(Fx^*) = 0$ since PF is closed, hence $x^* \in N(F)$ because $0 \in H$ is the unique zero of P. The uniqueness of $x^*$ is obvious.

Remark. Theorem 6 is a generalization of a main result established by K. Nickel for his BNM (see Theorem 1 in [1]).

It should be noted that in application the functionals u and v in (3.2)-(3.3) can both often taken to be positive and constant. In this case, the assumption (3.4) is naturally satisfied.

## 4. Specializations

In this section we specify the SRCA and its convergent properties to some concrete classes of nonlinear mappings.

### 4.1 Nickel's Class of Functions $\mathcal{F}$ and His BNM.

The class of functions $\mathcal{F}$, introduced by K.L. Nickel [1], is the set of mappings $f: B_0 \to R^n$ which satisfy the following property: For each set $C$, of the form $C = B(\bar{x},r) \cap B_0$ with $\bar{x} \in B_0$ and $r \geqq 0$, there exists a regular $n \times n$ matrix $\Lambda = \Lambda(C)$ and a real number $\lambda = \lambda(C)$ such that $0 \leqq \lambda < 1$ and for all $x,y \in C$

$$\|x - y - \Lambda(f(x) - f(y))\| \leqq \lambda \, \|\Lambda(f(x) - f(y))\| . \tag{4.1}$$

It is known that $\mathcal{F}$ is a subset of the Lipschitz bicontinuous mappings and for such a class Nickel established his BNM. We observe that every $f$ in $\mathcal{F}$ obviously satisfies the hypotheses (1.1)-(1.2) of Section 1 for the choice

$$u(x) = 1, \quad v(x) = \lambda, \quad \text{and} \quad P = \Lambda \tag{4.2}$$

in (3.2)-(3.3), and hence Nickel's BNM and his Theorem 1 on global convergence are proper specializations of the SRCA and Theorem 6 of this paper.

Based on the approach here, however, Nickel's class $\mathcal{F}$ can now clearly be amplified so that the BNM is applicable and convergence still holds. E.g., suppose that $N(f) \neq \emptyset$ and allow $\lambda \leqq 1$ and (4.1) holds just for all $y \in N(f)$, then (1.1)-(1.2) are also satisfied and hence the Theorems of Section 2-3 are all valid for the BNM.

### 4.2 Quasi-contractive Mappings and Williamson's Geometric Estimation Method.

A mapping $T: D \to H$ is said to be contractive if, there exists a positive constant $\alpha < 1$ such that for all $x,y \in D$

$$\|Tx - Ty\| \leqq \alpha \|x - y\| \qquad (4.3)$$

if the inequality holds just for all $y \in F(T)$, the fixed point set of T, we call such a T quasi-contractive mapping (in what follows, we always use "quasi" in the same way to indicate this restriction of y).

Let $\varepsilon = \alpha(1-\alpha^2)^{-1}$, $\delta = (1-\alpha^2)^{-1}$. T.E. Williamson, Jr. [3] has established the global estimation

$$F(T) \subset B(x - \delta(x - Tx), \varepsilon \|x - Tx\|) \quad \forall x \in D \qquad (4.4)$$

for a contractive mapping T, and in virtue of the estimation, designed a geometric estimation algorithm (abbreviate GEA) to construct a fixed point of T. Except for the difference of the choice of the initial point, his algorithm corresponds basically to the SRCA, namely when $u(x) = \delta$, $v(x) = \varepsilon$ in (3.2)-(3.3). But, his algorithm is not globally convergent. From the discussion here, cf. Theorem 6, apparently, the defect has been completely removed.

All conclusions for a contractive mapping can naturally extend to a quasi-contractive one, for the estimation (3.4) is really true for the latter. However, it is easily shown that the latter class of mappings is much larger than the first.

## 4.3 Quasi-strongly Monotone Mappings and the Authors' RCA.

A mapping $F: D \to H$ is said to be strongly monotone if, there exists a constant $\alpha > 0$ such that for all $x, y \in D$

$$(Fx - Fy, x - y) \geqq \alpha \|x - y\|^2 \qquad (4.5)$$

holds. We call the mapping F monotone if the inequality holds for $\alpha = 0$. For the equation $Fx = 0$, with F a quasi-strongly monotone mapping, we have really shown all of the convergence properties (except for Theorem 6) of the SRCA with the choice $u(x) = v(x) = (2\alpha)^{-1}$ and $P = I$ in (3.2)-(3.3). Especially, it is emphasized that, by the boundedness of a monotone mapping, the Bounded Convergence Theorem indicates that the SRCA is unconditional and always locally convergent for a finite and infinite dimension space, respectively.

In order to get the global convergence, we assume that F is also Lipschitzian, i.e., for some constant $L > 0$, $\|Fx - Fy\| \leq L\|x - y\|$, and specify the algorithm by

$$u(x) = \alpha^{-1} \quad \text{and} \quad v(x) = (\alpha^{-2} - L^{-2})^{\frac{1}{2}}.$$

Then the SRCA is globally convergent. To see this, notice that, $\lambda = (1-\alpha^2 L^{-2})^{\frac{1}{2}} < 1$, by the definition of $\lambda$, and every $x \in D$ and $y \in N(F)$, the quasi-strongly monotonicity and L-continuity of F gives that

$$\begin{aligned} \|x - \alpha L^{-2} Fx - y\|^2 &= \|x - y\|^2 - 2\alpha L^2 (Fx, x-y) + (\alpha L^{-2} \|Fx - Fy\|)^2 \\ &\leq (1 - 2\alpha^2 L^{-2} + (\alpha L^{-1})^2)\|x-y\|^2 = (1-\alpha^2 L^{-2}) \quad \|x-y\|^2 \end{aligned}$$

which shows that the mapping T defined by $Tx = x - \alpha L^{-2} Fx$ is really a contractive mapping with modulus of contractivity $\lambda$, so it immediately follows from 4.3 that

$$N(F) \subset B(x - g(x), r(x)),$$

i.e., the hypotheses (1.1)-(1.2) and assumptions of Theorem 6 are all satisfied.

## 4.4 Strictly Pseudo-contractive Mappings.

A mapping $G: D \to H$ is said to be strictly pseudo-contractive if, there is a positive constant $\beta < 1$ such that for all $x, y \in D$

$$\|Gx - Gy\|^2 \leq \|x - y\|^2 + \beta \, \|(x - Gx) - (y - Gy)\|^2 . \tag{4.6}$$

It is known [6] that G is strictly pseudo-contractive iff $F = I - G$ is monotone with the following property

$$(Fx - Fy, x - y) \geqq \frac{1}{2}(1 - \beta) \|Fx - Fy\|^2 . \tag{4.7}$$

So, we consider the latter class here, where $\frac{1}{2}(1-\beta)$ is replaced by $\alpha > 0$.

Let F be a monotone mapping with the property (4.7) and assume that F satisfies the following quasi-expansive condition

$$\|x - y\| \leq L\|Fx\| \qquad \forall\, x \in B_0,\quad y \in N(F). \tag{4.8}$$

We then easily show that the SRCA with the specialization

$$g(x) = \alpha^{-1}L^2Fx \text{ and } \quad R(x) = L^2((\alpha^2 - L^{-2}))^{\frac{1}{2}}\,\|Fx\|$$

is globally convergent (the reasoning is almost similar to the previous one).

Remark. Under the assumption that $\alpha < L < (\frac{1}{2}(1+5))^{\frac{1}{2}}\alpha$ in Subsection 4.3 and $\alpha < L \leq 2^{\frac{1}{2}}\alpha$ in Subsection 4.4, Wu and Wang [4] specify the SRCA by setting $u(x) = \alpha/L^2$, $v(x) = (\alpha^{-2} - L^{-2})^{\frac{1}{2}}$ , $P = I$ and $u(x) = \alpha$, $v(x) = ((L^2 - \alpha^2)^{\frac{1}{2}}$, $P = I$, respectively. Clearly, our specializations here not only abstain from their restrictions on $L$ and $\alpha$, but also increase the speed of convergence greatly.

Some more sophisticated specializations can also be done, for example, see [7].

## References


1. K.L. Nickel, A globally convergent ball Newton method, SIAM J. Numer. Anal. 18 (1981), 988-1003.

2. You Zhaoyong, Xu Zongben and Liu Kunkun, The region contraction algorithm for constructing zeros of quasi-strongly monotone operators, J. Engineering Math. Vol. 1 No. 1 (1984).

3. T.E. Williamson, Jr., Geometric estimation of fixed points of Lipschitzian mappings, II, J. Math. Anal. Appl., 62 (1978), 600-609.

4. Wu Yujiang and Wang Deren, On ball iteration method for a monotone operator, J. Engineering Math. (to appear).

5. J.M. Ortega and W.C. Rheinboldt, Iterative solution of nonlinear equations in several variables, Academic Press, New York, 1970.

6. F.E. Browder and W.V. Petryshyn, Construction of fixed points of nonlinear mappings in Hilbert space, J. Math. Anal. Appl. 20 (1967) 197-228.

7. Xu Zongben and Liu Kunkun, A application of the SRCA to the problem of constructive solvability of monotone mapping, to appear.

## Contributors


### Angelov, Rumen

High Institute for Economics

Varna

Bulgaria

### Engels, Hermann

Institut für Geometrie und
Praktische Mathematik
Technische Hochschule Aachen

Templergraben 55

5100 Aachen

West Germany

### Fischer, Herbert

Institut für Angewandte
Mathematik und Statistik
Technische Universität München

Arcisstr. 21

8000 München 2

West Germany

### Fujii, Yasuo

Educational Center for
Information Processing
Kyoto University

606 Kyoto

Japan

### Gardeñes, Ernest

Dpto. de Ecuaciones Funcionales
Facultad de Matematicas
Universidad de Barcelona

Gran Via, 585

Barcelona 7

Spain

### Garloff, Jürgen

Institut für Angew. Mathematik
Universität Freiburg

Hermann-Herder-Str. 10

7800 Freiburg i.Br.

West Germany

### Giec, Tadeusz

Institute of Mathematics
University of Łódź

ul. Tamka 4 m.31

91 403 Łódź

Poland

### Ichida, Kozo

Educational Center for
Information Processing
Kyoto University

606 Kyoto

Japan

### Kołacz, Henryk

Technical University of Poznań
Institute of Mathematics

Piotrowo 3a

60-965 Poznań

Poland

### Krawczyk, Rudolf

Bohlweg 2

3392 Clausthal-Zellerfeld

West-Germany

KRÜCKEBERG, FRITZ

Gesellschaft für Mathematik
und Datenverarbeitung
Schloss Birlinghoven

Postfach 1240

5205 St. Augustin 1

West Germany

LIU, KUNKUN

Department of Mathematics
Xi'an Jiaotong University

Xi'an, Shaanxi Province

People's Republic of China

MARKOV, SVETOSLAV

Bulgarian Academy of Sciences
Centre of Biology
Problem Group for Mathematical
Modelling in Biology

"Acad. G. Bončev" Str., Block 25

1113 Sofia

Bulgaria

AN MAY, D.

Rechenzentrum der Rhein.-Westf.
Techn. Hochschule

Seffenter Weg 23

5100 Aachen

West Germany

MIELGO, HONORIO

Facultad de Matematicas
Universidad de Barcelona

Gran Via, 585

Barcelona 7

Spain

MITROVIĆ, ŽARKO

Pedagosko-tehnicki fakultet
Đure Đakovića b.b.

Lenjinova 6/11

23000 Zrenjanin

Yugoslavia

NEUMAIER, ARNOLD

Institut für Angewandte Mathematik
Universität Freiburg

Hermann-Herder-Str. 10

7800 Freiburg i.Br.

West Germany

NICKEL, KARL

Institut für Angewandte Mathematik
Universität Freiburg

Hermann-Herder-Str. 10

7800 Freiburg i.Br.

West Germany

OZASA, MASAHIRO

Department of Electrical
Engineering
Ritsumeikan University

603 Kyoto

Japan

PETKOVIĆ, LJILJANA

Faculty of Mechanical Engineering

Beogradska 14

18 000 Niš

Yugoslavia

PETKOVIĆ, MIODRAG

Faculty of Electronic Engineering

Beogradska 14

18 000 Niš
Yugoslavia

RALL, LOUIS B.

Mathematics Research Center
University of Wisconsin-Madison
610 Walnut Street
Madison, Wisconsin 53705
USA

ROHN, JIŘI

Matematicko-fyzikálni
fakulta KU
Charles University
Malostranské nám. 25
118 00 Praha 1
Czechoslovakia

SCHMIDT, KLAUS D.

Seminar für Statistik
Universität Mannheim
A5
6800 Mannheim
West Germany

SHEN, ZUHE

Mathematics Department
Nanjing University
Nanjing
People's Republic of China

THIELER, PETER

Fachbereich Mathematik
und Naturwissenschaften
Fachhochschule Darmstadt
Schöfferstr. 3
6100 Darmstadt
West Germany

TREPAT, ALBERT

Facultad de Matematicas
Universidad de Barcelona
Gran Via, 585
Barcelona 7
Spain

XU, ZONGBEN

Department of Mathematics
Xi'an Jiaotong University
Xi'an, Shaanxi Province
People's Republic of China

YOU, ZHAOYONG

Department of Mathematics
Xi'an Jiaotong University
Xi'an, Shaanxi Province
People's Republic of China

[illegible]